BEYOND
SUPERNATURE

BANTAM NEW AGE BOOKS

This important imprint includes books in a variety of fields and disciplines and deals with the search for meaning, growth, and change. They are books that circumscribe our times and our future.

Lyall Watson
BEYOND SUPERNATURE

A New Natural History of the Supernatural

BANTAM BOOKS
TORONTO · NEW YORK · LONDON · SYDNEY · AUCKLAND

BEYOND SUPERNATURE

A Bantam Book / January 1988

*New Age and the accompanying figure design as well as the
statement "the search for meaning, growth, and change" are
trademarks of Bantam Books, Inc.*

Library of Congress Cataloging-in-Publication Data

Watson, Lyall.
  Beyond supernature.

  Bibliography: p. 269
  Includes index.
  1. Occult sciences. 2. Physical research.
I. Title.
BF1411.W33 1987   133   87-1348
ISBN 0-553-34456-0

Published in the United States

PRINTED IN THE UNITED STATES OF AMERICA

O   0 9 8 7 6 5 4 3 2 1

CONTENTS

v

CONTENTS

"Those who refuse to go beyond fact,
rarely go as far . . ."

THOMAS HENRY HUXLEY in *Collected Essays*, 1881

INTRODUCTION

As a biologist, I am fascinated by the soft edges of science, by the fleeting glimpses we get of strange shadows just beneath the surface of current understanding.

I tried in *Supernature* to redefine this fringe, to reconcile nature with what seems to be supernatural. And helped, up to a point, to create a sort of demilitarised zone into which both scientists and enthusiasts could go without abandoning either their sense of proportion or their sense of wonder.

But that was fifteen years ago and much has happened since. The publication of *Supernature* made me a focus for anomalous experience – and gave me the freedom to explore it at will. I have tried, along the way, to keep contact with those who share my excitement by putting out position papers in the form of six further books – each looking at the loose ends of the world in a slightly different way.

The time has come now, however, to go back to the beginning once again and see where we stand, almost a generation on.

During the last few years there has been a strong reaction against research into the unusual. Critics of anything paranormal have established influential committees with the express purpose of stopping such research altogether. They have succeeded, in at least one case, in destroying reputations by sending magicians, posing as psychics, to ingratiate themselves with a group of researchers, with the express intent of deceiving them at every opportunity. These tactics prove nothing, except perhaps a degree of intolerance which is blatantly unscientific. There are few fields which would be proof against such invasions.

Given wide public interest in the supernatural, it was probably inevitable that it should become a big business and suffer from all the distortions of the marketplace. I am ruefully aware of having helped to create this situation and accept my share of responsibility for fuelling enthusiasms which have, in some cases, got out of hand. Our culture, however, is prone to such excesses.

There is, when you look at it closely, no such thing as the supernatural. All we have are reports of experiences which seem to be beyond natural explanation – but we do have these in astonishing abundance. And the reports have become so frequent and so widespread that they are very difficult for anyone with real scientific curiosity to ignore.

I am fascinated by the fact that people all over the world have, and not just in our cultish time, come to accept the existence of some sort of paranormal reality. They hold beliefs in the existence of things such as spirits, of miraculous happenings, reincarnation, communication with the dead and telepathy amongst the living – and these beliefs are so persistent and so much alike that it is tempting to look for common cause.

Where do such ideas come from and what is it that sustains them, even in the face of official incredulity and scorn? Is it possible, even if the supernatural does not exist, that we need somehow to invent it?

I am not wedded to the proposition that the supernatural *must* exist. If one defines supernatural experience simply as – the experience of something unusual, something which exceeds the limits of what is deemed possible – then there is clearly a vast field of experience, of repeated experience, from all over the world, just waiting to be explored. The fact that such reports are, by their very nature, largely anecdotal, has led to their being discarded as unacceptable to science. Which is a pity and a waste, because I suspect that answers to some of the riddles of the paranormal might well lie in the pattern and content of such reports.

The greatest barrier to scientific acceptance of anything unusual remains its elusiveness. Which is a problem that leaves parapsychology – for the moment the most formal and least disreputable approach – an immature science without basic principles or consistent findings, hoping still to produce the elusive repeatable experiment. Failure so far to do so in the laboratory makes it easy for some orthodox scientists to dismiss the supernatural as meaningless; but it is difficult for anyone like myself, who has been involved in the field, that is outside the confines of the lab, to deny the common and powerful reality of experience that breaks the rules.

My own experience of the unusual in action in a wide variety of cultures, suggests very strongly that there is something well worth pursuing. I have watched the rise of interest in the occult – and the inevitable backlash – with fascination. I have shared the high expectations of those trying to get to scientific grips with telepaths and metal-benders; and suffered with them the disappointment of discovery that the phenomena are strangely, almost wilfully, elusive. I understand the disillusion which has resulted, but must say that nothing has happened in the last fifteen years to alter my certainty that we stand to learn important things about ourselves from scrutiny of those areas in our lives that can be almost commonplace, but nevertheless defy easy description.

I believe that what the supernatural very badly needs is a new and fresh and thorough overview. A cross-cultural survey of the paranormal. An ethnography of the unusual. A broadly based and well-funded professional operation designed to retrieve and catalogue and classify all unusual events everywhere.

This is unfortunately not it. This is nothing more than my own personal attempt to make sense of what I have seen and heard in recent years. It is an attempt to define and describe the range of unusual experience a little more precisely. An attempt which I sincerely believe to be necessary, because I remain convinced that there are things going on around us which cannot easily be squeezed into forms that fit the accepted mould.

So, despite the cavils of self-appointed committees for the suppression of curiosity, I continue to pursue ghosts on the edges of perception. I persist in pointing out inconsistencies in natural history – not because these necessarily mean anything in themselves, but because they could lead to better understanding of what is usual through a new and more open-minded analysis of the pieces that don't quite fit.

And as with *Supernature*, I offer this new survey to all those who can still look at the world with wide eyes – and wonder.

Lyall Watson
Ballydehob, Ireland; 1985

Part One

LIFE

"There is one common flow, one common breathing,
All things are in sympathy."

HIPPOCRATES in *De Alimento*, Fifth Century BC

The life sciences are in a curious state.

Ever since the discovery of the structure of DNA in 1953, they have been dominated by molecular biology. We have cracked the genetic code which determines the sequence of amino acids in proteins. We know most of the details involved in protein synthesis from these acids. We have started to unravel the mysteries of those special proteins known as enzymes which knit assorted biochemicals into the complex machinery of viruses and bacteria. And we begin to understand, in principle at least, the astounding regulations which govern the workings of a living cell.

This prowess, rewarded and reinforced by several Nobel Prizes, has produced a kind of academic euphoria – a feeling that, at last, we are on the brink of a full explanation of all the phenomena of life.

Nothing could be further from the truth.

Our ability to ask, and answer, questions about the mechanics of life has drawn attention away from our continuing inability to understand the real nature of living things. Our impressive achievement at physical and chemical levels, conceals an almost total lack of progress in coming to terms with general biology.

We know a great deal about the parts of living things, but next to nothing about the process which assembles those components into a functional whole.

All life possesses properties which are peculiar, which cannot be understood in terms of the properties of the isolated parts. The whole creature is always much more than the sum of its parts. It contains structures and exhibits behaviour which cannot be predicted from a study only of the known ingredients. There is something missing from the mechanistic model, something which

7

seems to have no roots in even the most sophisticated biophysics or biochemistry.

Life remains mysterious. Biology is rife with unsolved, perhaps insoluble, problems.

Classical physics was revolutionised in 1927 by Werner Heisenberg's uncertainty principle, which made it clear that certain microphysical events could never be completely known. That, despite our best intentions – to a very real extent, directly *because* of these intentions – some things could be predicted only in terms of probabilities. Biology has yet to produce its Heisenberg, yet to come to terms with such weakening of the traditional laws of cause and effect.

In this first section, I want to look at a few of the consequences of this imbalance in understanding. And show how it might begin to be restored by concentrating more on the form and shape of whole organisms than on the details of their structure.

There are certainly some astonishing things going on.

Chapter One

PATTERN

Consider the camel.

Somebody should. Much maligned as "a horse designed by a committee" or as a bad-tempered beast with an evil smell, the camel is wonderfully well-adapted to its environment. With huge widely splayed feet equally suited to soft sand or snow, stomachs with elaborate extension sacs for holding extra water, humps full of excess energy, lips tough enough to deal with the most thorny scrub, and eyes protected from windblown debris by devastating lashes – no large creature is better suited to deal with the world's deserts. But perhaps the most extraordinary item in this catalogue of special features, is one that is seldom considered at all. Camels have peculiar knees.

When a camel lies down, as camels frequently do in order to get out of the wind, or simply to demonstrate their loathing for all forms of labour, they fold at the knee. The forelegs crumple like origami, backward at the top and forward lower down, so that a resting camel lies with its feet tucked up beneath it, leaning on the long shaft of the lower leg. It is hardly surprising that constant abrasion at both ends of this bone should leave camels with tough callosities at the joints. But what is surprising is that these calluses don't have to be acquired. Baby camels are born with them.

And that presents a problem.

The currently accepted theory of evolution makes a number of assumptions. The first is that changes in living things are the result of random mutations in their genes. The second is that these changes are subject to the pressures of natural selection in which harmful ones are weeded out and abandoned. And the third is that the inheritance of useful changes is subject only to the laws of genetics. In other words, the genes for camels' knees are prey to the usual influences and, from time to time, undergo rare and random alteration. Most changes of this kind are bad, inclined to upset the balance of the camel and disappear before they get the chance to be passed on. But by sheer luck, once in a very long while, a change can be beneficial – adding extra horny layers of skin to a poorly protected joint – producing individual camels which have a selective advantage and survive at the expense of their fellows with weaker knees.

This is Charles Darwin's hypothesis, later fortified by the genetic discoveries of Gregor Mendel, and now known as the neo-Darwinian theory of evolution. It is a useful and plausible scenario, so well-accepted now that it is the only one you are likely to find in many textbooks – despite the fact that there is no direct evidence to support it, and that it fails to account for everything that happens.

The problem with the calluses on camels' knees is that they are inherited. They are visible on a baby camel long before it has ever had to kneel on stony ground. If the vital thickening in the skin occurred after birth, as a result of constant friction, there would be no problem. But it begins to take place in the embryo at precisely those places the little camel's ancestors found to be vulnerable and in need of protection. It is hard to avoid the conclusion that the calluses represent a specific response to a particular environmental pressure experienced by the camel's ancestors. And it is impossible to explain this response in terms of orthodox evolution. But the neo-Darwinian theory is not the only one available.

In 1809, half a century before Charles Darwin published his *Origin of Species*, a French naturalist drew attention to the fact that species change and develop. Jean Lamarck believed that the experience of parents was not necessarily wasted, but could provide direct benefits to their offspring. He suggested that progressive changes which were related to the vital needs of a species might be inherited.

Lamarckism, the theory of the inheritance of acquired characteristics, was once very popular. Darwin himself espoused it. But by the turn of the century, it led to violent controversy. It is difficult at this distance to appreciate the passions that were aroused. Those who believed that changes in living things took place, at least in part, in response to circumstances and experience – were ranged

10

against those who insisted that changes were random and subject, only after they had been made, to the pressures of selection. It was a battle between those who thought that there was something qualitatively different about living things – and those who were coming to believe that everything could be reduced to physics and chemistry.

The arguments ended rather suddenly with the rediscovery of Mendel's work in genetics, which provided an elegant mechanism for natural selection to operate by chance, and gave the advantage to the Darwinians. So complete was their victory, that Lamarck's ideas were branded as "ignorant, superstitious and disreputable" and declared off-limits to serious students of biology.[72] They still are, but there are signs of a neo-Lamarckian revival.

Faced with the Curious Case of the Camel's Knee, and the problems it poses for Darwinism, it is worth going back to what Lamarck actually said about the effects of experience on evolution. He said, "alteration in the environment of animals leads to alteration of their needs, which in turn produces modification in their structure and organisation."[218] In camel terms, this means that as the grasslands turned to desert and the animals trapped there needed to escape the new extremes of heat and cold, any structural change which suited their needs in this new way of life – such as knee pads which make lying down on rough ground more comfortable – would have been more likely to be selected.

Does this sound ignorant or superstitious? Hardly. It seems to make good sense and, on the face of it, offers no great violence to current evolutionary theory. But it does in fact represent a radically different approach. It allows evolution to be seen as a cumulative and responsive process, instead of one that is entirely accidental. It introduces the possibility of purposiveness – an idea that evolution may be more than just a random and desperate struggle for survival. And that, in the current climate of science, is certainly heretical.

It is a heresy I am glad to defend. I am impressed by the strengths of the neo-Darwinian theory. Its triumph has been to explain how change takes place and species diverge. But there is nothing in it to account for developments which are not only increasingly complex, but wonderfully relevant. It is difficult to see how random mutation could, in a comparatively short time, and without reference to the environment and experience of previous generations, put calluses on camels' knees in precisely those places they need them.

Camels' knees may seem like small things in the great march of life on earth, but their appropriateness is perplexing. And they are not alone. Warthogs, which frequently feed on their knees, have similar, equally well-placed, horny callosities. Ostriches inherit bulbous thickenings on the underside of their bony sternums, placed fore and aft, precisely where they sit. The skin on the soles of our feet is much thicker than it is anywhere else on our bodies – and it becomes so in the womb, long before our bare feet ever touch the ground.

Accidental mutation and the rigours of natural selection over long periods of time, don't seem to be sufficient explanation. It is difficult to avoid the conclusion that, in addition to the neo-Darwinian mechanisms, there is some other more creative influence at work in evolution. A force which lets things move and grow in ways that give them meaning.

A lot of what happens may well be due to chance, but there are some interesting patterns superimposed on this basically random world.

Chance

Most things happen by chance, which is not the same thing as saying that they happen by accident. Chance seems to have a pattern and a reason of its own.

After a lifetime of research in physics, Erwin Schrödinger concluded that "in the overwhelming majority of phenomena whose regularity and invariability have led to the formulation of the postulate of causality, the common element underlying the consistency observed – is chance."[332] In other words, nature is governed by the laws of chance.

Einstein was uncomfortable with the idea. "I cannot believe," he said, the year before he died, "that God plays dice with the cosmos." He found it easier to accept that there were no laws at all, that the universe was in a state of chaos, than to believe that the world was run by statistics that compelled "the good Lord to throw the dice in each individual case".[93]

Now, thirty years on, it begins to look as though we may indeed be part of some great game – an affair which follows a set of cosmic rules.

The word "chance" comes from the Old French *cheance* meaning "the way in which things fall", and seems, right from the start, to have referred to the lie of gaming dice – the oldest, and still one of the most effective, randomising devices. The French word *cheance*, in its turn, derives from the Latin *cadentia*, which also means "falling", but carries in addition, a sense of rhythm. Hence the English word "cadence" that describes a measured movement, something pleasantly musical. This etymology implies that chance was never seen as haphazard or altogether neutral. It hints at an early awareness of the existence of some kind of pattern, at consciousness of an accent in the rise and fall of the dice, at the presence or absence of what we have come to call luck. It is something that echoes through history like a pulse beat.[58]

Cadence in the fall of early dice was ascribed not so much to luck as to providence, to the will of the gods. And in the beginning, attempts to influence the fall were largely votive, concentrating on ways and means of catching the attention of the deities thought to be responsible. Astragalomancy, the art of contact with such gods, of divination by dice, is variously attributed to the Lydians in Asia Minor in the sixth century BC, or to Palamedes, a son of the King of Euboea, who introduced it to Athens during the fourth century BC. But it is certainly older than that. Long before dice became formal cubes of wood or stone, they were actual *astragali*, roughly cubical bones taken from the ankles of antelope or sheep. Such bones, some of them marked with appropriate symbols, have been found on prehistoric sites in Africa dating back over 40,000 years. And their use seems soon to have become almost universal.[156]

There are forms of dice in African, Inuit and Mayan ritual. Siva, the Hindu god of a thousand names, when he isn't dancing or destroying, is shown determining the fate of mankind by throwing dice. In old Tibet, the Jalno, personal representative of the Grand Lama, regularly played dice with the national scapegoat – and always won, because he played with loaded cubes. And even the Bible allows that "If there is need, it is lawful with all due reverence to seek the judgement of God by lots." Meaning, when in doubt, try dice.

One of the best-known modern demonstrations of the way in which things fall, is the game of chance called craps. The rules are very simple. It is played with two dice and the person who throws them wins immediately if the total of the spots on the first throw is 7 or 11 – or loses immediately if the total is 2, 3 or 12. If the total is 4, 5, 6, 8, 9 or 10 – the dice are thrown again, and continue to be thrown, until the player "makes the point" by duplicating the total of the first throw; or "craps out" by throwing a total of 7 and is forced to give up the dice to another player.

The probability of "making the point" is roughly 49 per cent – odds against the thrower by a very small margin, just enough to give the house an edge in the long run. These odds remain the same for each individual throw, but those against any thrower winning several times in succession, rapidly become astronomical. And yet it happens. In 1950, an unidentified man walked up to a crap table at the Desert Inn in Las Vegas and successfully held the dice for over 80 minutes. During this time, he made his point an astonishing twenty-eight consecutive times – and cost the casino a great deal of money.[400] Someone wins, someone loses.

Sometimes, everybody loses. On August 24th, 1983, the Amtrak Silver Meteor train made its usual journey from Miami to New York, but the trip turned out to be anything but routine. At 7.40 p.m. in Savannah, Georgia, the train struck and killed a woman who was fishing from a bridge. At 9.30 p.m., just sixteen miles further on, it hit and destroyed a truck that was parked too close to the tracks in Ridgeland, South Carolina.

The shaken crew were replaced, but at 1.10 a.m. the following morning, the train hit a tractor-trailer on a crossing at Rowland, North Carolina and two passenger cars were derailed, sending 21 people to hospital – including the engine driver. Once again the crew were replaced and the train travelled on, but at 2.37 a.m. in Kenly, North Carolina it ran headlong into a car that ignored the warning lights at yet another crossing. And the National Transport Safety Board, though they had no reason to doubt the quality of the machinery or to question the skill of the crews, declared Amtrak 117 a "Rogue Train" and cancelled the rest of its journey.[304]

Runs of good and bad luck do occur. Events, which ought to be distinct, sometimes have a tendency to gather in clusters which surprise us. And when random occurrences arrange themselves in such apparently purposeful ways, we tend to look at them rather hard, massaging them for meaning, forgetting for the moment that in the final analysis, in the long run, they almost have to happen.

This is, however, a very recent and still somewhat restricted understanding. When one looks back at the history of gambling, it is clear that no game of chance was ever seen as just a game or left entirely to chance. Every such game ever played, has gathered ceremonial aspects which suggest attempts to control or influence its destiny. Every gambler, ever since the games began, has looked for some way of gaining a personal edge. And when prayer and superstition, when appeals to Lady Luck, were not enough, they turned to the new tools of science.[90]

By the sixteenth century, the rudiments of classical science and the rigours of Greek scepticism had taken sufficient hold of the Italian marketplace, for the Renaissance to begin. A new independence was abroad and in this climate of enquiry, it seemed logical for a group of gamblers in Pisa to turn for help with their particular gaming problems to the greatest scientist of their day. And Galileo saw nothing incongruous in time spent, day after day, down on his knees in a dice game, puzzling over patterns in the fall of the cubes. The result of his exertions was greater success for the gamers as a result of a new set of formulae which, for the very first time, tamed chance by laying down the laws of the game.[122]

A century later in France, the young mathematicians Blaise Pascal and Pierre Fermat, also commissioned by a dice player down on his luck, turned these formulae into their Theory of Probability, further reducing the vagaries of chance to a very elegant grammar and arithmetic. "If something is certain to happen," they said, "let its probability be 1. And if it is certain *not* to happen, if it is impossible, let its probability be 0." Anything else, which in practice means almost everything, has a probability which is expressed as a fraction between these limits. And if the probability is less than half, the odds are against that event taking place.[274]

The beauty of this system is that it eliminates any discussion of improbability. With it, we can get useful and reliable information from things that are in themselves uncertain. We remain unable to predict the fall of a single die, because there is an equal chance of any side coming to rest face up, and there is nothing, either in the pattern of previous throws or in the context of the game under normal circumstances, which can change that. But with the theory of probability, we can begin to make accurate predictions about the outcome of a vast number of falling dice – or about the workings of the universe. It seems a little odd, at first glance, that we should be able to predict the movement of great planets successfully, while remaining unable to control or foretell the fall of a tiny die. But, as the scientific philosopher Sir Karl Popper points out, "In order to deduce predictions one needs laws and initial conditions; if no suitable laws are available or if the initial conditions cannot be ascertained, the scientific way of predicting breaks down."[280] And with dice, we clearly lack sufficient knowledge of initial conditions. A rolling die passes through a number of states in which even a minute disturbance will affect its further course, making the result dependent in the end, particularly with a perfect die, on the conditions of the atoms at its edges. And these, in accordance with the uncertainty principle, are impossible to predict.

And yet there are gamblers, a small number of inveterate and successful rollers of dice, who seem to be able to do so – at least often enough to make an idle living.

Science has problems with them. It doesn't recognise luck, dismissing it out of hand as a "superstitious misinterpretation of the idea of chance". It may indeed be nonsense to suggest that each of us is dogged by a mysterious influence that is attached to us like a shadow, but – in the words of a well-known oil millionaire, "I'd rather be lucky than smart, 'cause a lot of smart people ain't eatin' regular."[7]

The closest we have come, perhaps, to defining luck is a look at its negative side, at those who seem to have less, rather than more, of their fair share of it. In English, we use the word "accident" to describe both chance events and those that cause disruption, as though there were no distinction between them. German, as usual, is more precise – distinguishing between *der Zufall*, which is a truly unforeseen course of events, and *der Unfall*, which is a mishap involving damage or injury. The first event is a random one, and recognised as such. The second one may not be.

There seem to be some people and some situations which are unpredictably predictable. The ones sometimes described as "accidents looking for a place to happen". This concept of "accident-proneness" was introduced to psychology in 1926, and has been most rigorously tested on the drivers of motor vehicles.[108] A survey of bus drivers in Northern Ireland, for instance, after eliminating all possible mechanical and environmental influences in the causation of accidents, concluded that there was a personal factor involved. In other words, even when exposed to equal risk, some individuals were more likely than others to have accidents. It seems that, within their lives, these unhappy individuals have "spells" during which they are predictably prone to crash.[68]

There has been only limited success in attempts to describe psychological or physiological characteristics which would make it possible to recognise such individuals or their times of greatest risk far enough in advance to take them off the road. But insurance companies have isolated what they claim to be a direct statistical connection between occupation and accident. In Britain, school teachers, bakers, bank officials, local government employees and off-duty police, prison and fire officers are seen as low risks on the road. While students, doctors, butchers, bookmakers, television producers, professional entertainers and all members of the armed forces, either on or off duty, are considered to be high risks and burdened with appropriately loaded premiums.[60]

The use of the word "consider" in this context is an appropriate one. It began as an astrological term, dealing with the influence of the stars (*sidera*) on a contemplated decision. Actuaries today seem to earn their salaries, the decisions they make appear to save their companies money, but the rationale which puts an individual butcher at greater risk than a baker is hard to define or justify on any objective ground – except the Law of Large Numbers, which ensures that the customers take all the risks and the companies never lose.

The theory of luck is not alien to biology. In evolutionary terms, a belief in good or bad luck makes excellent sense. It is clearly beneficial, providing a useful social balm, protecting the successful from a certain amount of envy on the part of their peers. If luck is all it takes, success could come to anyone! The successful themselves, however, tend to be less convinced. Alexander Fleming's response to those who remarked on the "accident" of his discovery of penicillin, was that "chance favours the prepared mind."[59]

There's a lot in that. And it might be said, with equal validity, that "mischance follows the unprepared or unwary mind." But is that really all there is to it? Can fortune be nothing more than an ability, a willingness, to take the chances offered? And do we all get the same chances? The evidence suggests that we don't. The lives of individuals are just not long enough to let them play with planets or enjoy the actuarial benefits of sufficiently large numbers. Some of us, in the natural course of things, are bound to slip between peaks in the cycle of fortune.

We have, however, the right to expect a certain number of lucky breaks. And when these occur unusually often, or fail to happen at all, it is not unreasonable to wonder whether the laws of chance are being, if not broken, at least a little bent by the action of some other influence.

Coincidence

In 1967, the telephone number of a London police station was changed to 40116. An officer working there asked an acquaintance to call him during duty hours the following evening, but told this person that the number was 40166. It was not until he came to work the next day, that the policeman realised he had got the number wrong. That evening, however, while patrolling an industrial estate with a colleague, the officer noticed a light burning in one of the factories and went in to investigate. Just as they arrived, the telephone rang. The policeman picked it up and was astonished to find that the caller was the person to whom he had given the wrong number. There was no number on the instrument, which turned out to be ex-directory, but the factory manager confirmed later that it was in fact 40166.[213]

Arthur Koestler described such coincidences as "puns of destiny" and suggested that the hint of purpose in them made these qualitatively different from the purely mathematical long shots that, for example, from time to time give bridge players the one hand in 635, 013, 559, 600 that contains all thirteen cards of a single suit. Such combinations of cards are, according to the laws of chance, as likely as any other hand. They are certain to happen. The odds against them are probably no greater than those which let a policeman's acquaintance ring him at the precise moment the officer comes within reach of the one telephone in his district that will ring in response to the wrong number. But there is something in "meaningful coincidences" of this nature that ought to make a biologist sit up and take notice – particularly when they appear to have survival value.

There is a Victorian story of a country girl who set out to visit her sister Elizabeth Mary Parker, who lived at 36 Eaton Place in London. En route she lost her wits, her memory and her way and knocked in desperation at the door of a house number 36 in a street that seemed to have the right name. It turned out that someone called Elizabeth Mary Parker did in fact live there and was not her sister, but was able, because of earlier mix-ups in the mail, to tell the distraught girl where her namesake, the true sister, could be found.

Coincidences of this order are both entertaining and instructive. They draw attention to the tendency of what the Greek physician Hippocrates called "sympathetic elements" to seek each other out. And they hint at a mechanism behind the laws of nature.

Think for a moment of the dilemma faced by someone trying to find a friend in a city the size of London. The odds against running into them by accident are enormous. A random walk through the streets, looking at the names on doorbells, might be successful – but it would take far too long. The only sensible procedure is one which eliminates the largest number of alternatives. The fact that the Victorian girl knew the number 36 was a start, narrowing down the list of choices. A street name would have been better, but as it happened, the combination of number with a personal name and a little luck, was sufficient.

Evolution seems to follow a similar course. The number of conceivable molecular structures for even a single protein is so large, that they cannot even be screened. And yet, somehow, the process of natural selection succeeds in moving from the general to the specific, in eliminating alternatives and putting together the right ingredients for stable life, in a remarkably short time. Part of this process is lawful and orderly in the sense that choices are made according to established rules, under which each new decision depends on the answer to the previous question. This leads inevitably towards greater stability, but there is room and perhaps even a need for a random and unstable ingredient. For an element of luck. It was coincidences of this kind that started Austrian biologist Paul Kammerer thinking about clusters of like events that seemed to be connected by a common principle.

Kammerer was "the Linnaeus of coincidence".[211] He spent his spare time in public places, recording and classifying everything that happened. He noted the number of people passing, grouping them according to how they dressed and what they carried, finding

19

clusters of phenomena wherever he looked. He arranged occurrences into morphological series, distinguishing between coincidences of the first, second and third order, and organising events in accordance with the number of attributes or parameters they shared. He concluded that incidents were patterned in time and space, recurring in ways which could not be explained by coincidence alone. Things happen, he said, in ways "which make coincidence rule to such an extent that the concept of coincidence itself is negated". In other words, it never rains but it pours. And he called this tendency, the Law of Seriality.[203]

It is not a law that depends on any causal forces we can measure, but there is something familiar about it that is very persuasive. Einstein found it "original and by no means absurd". There are echoes of it in Plato and Pythagoras, and in the cosmologies of Johann Kepler and Sir Arthur Eddington. All suspected the existence of some principle that transcends cause-and-effect, some kind of form, a sort of harmony that chimes through the universe, preceding and shaping the content of things.

The best attempt to lay this ghost in the machinery of things, is perhaps the collaboration in 1952 of a Nobel Prize-winning physicist and a great psychologist. Wolfgang Pauli and Carl Gustav Jung attributed coincidence to an acausal connecting principle which operates independently of the known laws of classical physics. They proposed a force at work in the world which tends to impose its own kind of discipline on the chaos required by the second law of thermodynamics. Natural law, they pointed out, is not absolute. It is a statistical truth which holds good only on average and often bears little relation to the way in which things actually happen. The law leaves room for exceptions, which can be completely acausal, but are nonetheless real and evident. And when such exceptions operate independently of time and space, linking events in strange and unexpected ways, meaningful coincidences take place. They called this Synchronicity.

Jung made it clear that he was aware of the consequences of their theory. If causeless events exist, if they cannot be ascribed to any known antecedents, then they must be considered as original creative acts. "Though we must of course guard," he said, "against thinking of every event whose cause is unknown as causeless. This is admissible only when a cause is not even thinkable." Many meaningful coincidences could be the result of pure chance. This is a thinkable explanation. "But," he went on, "the more they multiply and the greater and more exact the correspondence is, the more their probability sinks and their unthinkability increases, until they can no longer be regarded as pure chance but, for lack of a causal explanation, have to be thought of as meaningful arrangements."[200]

It is the nature of such arrangements, and the apparent purposiveness of many of them, that makes the possibility of coincidence something of real concern for biology.

In his Gifford Lectures twenty years ago, the zoologist Sir Alister Hardy drew attention to the truly creative power of natural selection. He discussed a variety of adaptive colour markings in animals and showed how many of these, particularly those which produce disruptive or cryptic patterns that make an individual hard to see, could not possibly have been produced by the organisms themselves. These camouflage markings show such subtlety of shade, contrast, tone value and position – all visible only at some distance from the animal concerned – that they can only have been selected by outside agents. The agents, of course, were predators who came amongst them like voracious art critics, eating all the less successful productions.[147]

Jacques Monod, geneticist and arch-mechanist, has no problems with such things. He insists that natural selection, without any outside help, is perfectly capable of drawing "all the music of the biosphere" from sources of noise. And that genetic accidents, occurring purely by chance, are sufficient explanation for the emergence of all complex structures and even the most purposive forms of behaviour. "Chance alone," he says, "is at the source of every innovation, of all creation in the biosphere."[248]

There are few field biologists able to share such presumption.

During the past decade, I have had the pleasure of spending time in the Amazon and on the edge of rainforest in the Mato Grosso of Brazil. It is a wonderland with biological variety on display in such abundance, it is almost embarrassing. Amongst it all, nothing recently seen has amazed and intrigued me more than the structure and behaviour of the giant anteater *Myrmecophaga tridactyla*.

This enormous creature is one of nature's ultra-specialists. It eats only ants and termites, ripping open nests and mounds with the aid of claws like sickles on the end of forearms which would not look out of place on a blacksmith. Its ability to rise up on hindlegs

21

and strike out with great speed, makes it one of the least-molested of all mammals, able to hold its own even against a jaguar. And yet, it has developed one of the most intricate of all patterns of disruptive colouration.

Across the anteater's icicle-shaped snout is a shadow patch which destroys its sharp profile as effectively as any gun camouflage pattern ever painted on a piece of field artillery. And on the forelimbs are bands of black and white fur which look randomly placed on an animal in motion, but line up perfectly to produce coincident stripes of camouflage across the front of an anteater resting on its haunches. The continuity of these colour bands when the legs are held together, is complete, a miracle of selective genetics as impressive as those which give some butter-flies and moths a continuous "design" that overlaps from one wing to another when these are held at rest.

I am prepared to believe that natural selection, given sufficient time, has the creative power to make such alignments. But on my last visit to South America, I became aware of one further order of coincidence amongst anteaters that puts a great strain on such credulity.

In addition to the leg and snout bands, giant anteater are emblazoned with a dark and diagonal side stripe, a vivid black slash that begins broad at the chest and tapers to a point over the pelvis. This, once again, breaks up the natural outline of the beast, which is already rather bizarre, and makes it very difficult to spot against a broken background. Young anteaters are similarly marked and enjoy the same protection as their parents, but what astonished me beyond measure was what happened when I watched the two come together. Baby anteaters ride on their mothers' backs, lying flat across the spine low down near the tail, clinging to the one spot there that makes the stripe on their own flank a perfect extension of the maternal sash. And as they grow larger, and move further forward to distribute their weight more evenly, their growing stripes continue to make a seamless connection, joining mother and child invisibly by means of the only shape in the only place that would make such continuity possible.

Mutations which produce a coincidence in pattern between different parts of the same animal, are impressive. Coincidence in the pattern between two animals, particularly when this depends on an effect that neither can see, but must be created by an

22

adaptive pattern of behaviour that has to change with age to be of any use at all, is awesome. And given the fact that it occurs in a species that does not seem to be under particularly heavy or critical predation, I have trouble accepting that what I see there is entirely the result of blind chance or happy accident. I am driven, more and more, to the conclusion that, in such examples at least, natural selection is joined by some other, more purposive, force.

Order

In the world of classical physics, there was no room for chance or coincidence. Everything had a place and a direct cause. Isaac Newton believed that it was just a matter of time before all such causes were known. "If it were possible," he said, "to know the position and velocity of every particle in the universe, then we could predict with utter precision the future of those particles – and therefore, the future of the universe."[268]

Heisenberg changed all that. He showed that it was impossible even to understand the present in all its detail, because we can never know the position *and* momentum of even one particle precisely. So in the world of modern physics, nothing has a cause. Things just happen, by chance, for no particular reason. But the way in which these happenings group themselves, is anything but random.

Certain substances, for instance, are radioactive. They spontaneously disintegrate, throwing off unstable particles in ways which are totally uninfluenced by temperature, pressure, chemical change or the effects of electric or magnetic fields. The disintegration is observable – as the wayward particles spin off, they produce a visible trace on a fluorescent screen – but it is a process that cannot in any way, even in theory, be predicted in advance. Schrödinger puts it well: "If you are given a single atom, its probable lifetime is much less certain than that of a healthy sparrow. Indeed, nothing more can be said about it than this: as long as it lives, the chance of it blowing up in the next second, remains the same."[333]

The future of any such atom has no relation to its past history or to its present environment. It is, in the words of physicist David Bohm, "completely arbitrary in the sense that it has no relationship whatsoever to anything else that exists in the world or that ever has existed".[26] And yet there is a pattern to such decay. The time it takes for half the atoms in a radioactive substance to disintegrate, is fixed and well-known. It is called the *half-life* of an isotope and varies from a millionth of a second to more than a million years. The half-life of radon is 3.825 days and of radium 1622 years. While it takes 4.51 billion years of uranium to decay to half its original level.

So, despite the fact that the disintegration of any one atom is totally unpredictable, the behaviour of large numbers of such atoms is assured. If disorder is sufficiently widespread, it becomes orderly. If you pile uncertainties up high enough, the outcome is certain. Random events, provided there are enough of them, produce patterns with reason and meaning. This is the paradox of probability – the conjuring of certainty out of collections of random and unrelated events. It is something governed by what the Swiss mathematician Daniel Bernoulli has called the Law of Large Numbers.

Put very simply, Bernoulli's Law says that "probability tends towards certainty as the number of events involved approaches infinity."[75] For example, if two puppies share a single flea, only one dog can have it at a time and there is no way to predict where the flea will be. But over a period of weeks in the den, it becomes increasingly likely that the dogs will share an equal number of bites. Or, to take another commonplace example, in the insurance business it is the customer who assumes almost all the risk. The man who buys a policy cannot know whether it will ever benefit him. But the insurance company, by basing its calculations on the Law of Large Numbers, on statistics derived from many thousands of similar policies, can guarantee that it will always make a profit.

The Law works so well that when it appears to fail, there is good reason to look for alternative explanations. The statistician Horace Levinson analyses two experiments with dice involving 49,152 and 315,672 throws respectively. In the first, a success was recorded when either 4, 5 or 6 came up – and in the second and longer series, success was limited to the appearance of only those sides with 5 or 6 spots. The Law predicts that half of the first series, that is 24,576 throws, should have been successful. The actual result was 25,145 – an excess of 569. The theoretical prediction for the longer series was one third, which is 105,224 successful throws. And in this case the result was 107,665 – an excess of 2441. In each series, the unexpected excess was close to 2.3 per cent, a result which could have been expected on the basis of chance alone, only once in five million times. Levinson concludes that the dice involved, perhaps even all dice, have a very small bias due to the different number of spots, which shows up only in very long series of trials.[213]

He could be right. Or, if the dice were not being thrown mechanically, the distortion could be evidence of the intrusion of some sort of human factor – which is not always easy to recognise.

I know an ecologist who, as a rule of thumb, assumes that if in a strange land he finds four trees of one species in a straight line, humans were probably involved in their cultivation. And if, on measurement, he discovers that these trees are equidistant, he regards his assumption as a certainty.

Most of the time, he is justified in doing so. Such regularity in small numbers is rare in nature. But there are occasions when nature is capable, without our help, of astonishing feats of order – even without resorting to the smoothing Swiss effects of Bernoulli's Law. It relies instead, in these cases, on an Italian influence.

The mathematician Leonardo Fibonacci lived in Pisa during the twelfth century, when it was a great mercantile centre with strong commercial ties to North Africa. He was tutored by an Algerian, learned the Arabic system of arithmetic notation and published an important book in which he also posed the following problem:

A certain man put a pair of rabbits in a place surrounded on all sides by a wall. If it is supposed that each pair begets a new pair every month, beginning in the second month, how many rabbits will there be in a year?

The answer is that 1 pair is born in the second month, 2 pairs in the third, 3 pairs in the fourth, 5 in the fifth, 8 in the sixth, 13 in the seventh, 21 in the eighth, 34 in the ninth – and so on. In other words, the number of pairs produced in each successive month is equal to the sum of the pairs produced in the two previous months.[100]

This sequence is now known as the Fibonacci Series and its pattern is by no means arbitrary.

If you take the common daisy *Bellis perennis* and look closely at its golden pin-cushion disc, you will find that it consists of dozens of

tiny, tightly-packed florets. These are never arranged haphaz-
ardly, but in a double set of curved lines spiralling out from the
centre. There are precisely 21 florets in the clockwise spiral and 34
on the anti-clockwise curve. In other composites – such as mari-
golds, thistles and sunflowers – the ratios range from 13/34 through
21/55 to 34/89 and on up, depending on the size of the flower head.
But all the numbers are always Fibonacci numbers.

In other plant families, the angle of divergence, that angle which
lies between the midline of successive leaves on the stalk, follows
the same sequence. Expressed as fractions of a circle, these are 1/2
in grasses, 1/3 in sedges, 2/5 in roses, 3/8 in bananas, 5/13 in leeks
and 8/21 in pine cones. Once again the Fibonacci connection holds.
Both the numerators and denominators of these particular
fractions represent the sum of the two previous numbers of each.

Even within a single species, such as the waterweed known as
quillwort *Isoetes lacustris*, a Fibonacci series is evident in changes
that take place with increase in age. The first leaves are diametri-
cally opposite each other and, as others are added, the fraction
changes to 1/3, 2/5, 3/8, 5/13 and so on until the limit of divergence
is reached.

It is difficult to dismiss such arrangements as coincidental. There
are similar progressions in the structure of antelope horns and in
the genealogy of the male honey bee. There are even said to be
relationships between the Fibonacci Series and the fabled Golden
Mean. The pattern is so pervasive that in California (where else?) a
Fibonacci Association was formed in 1962 "to exchange ideas and
stimulate research on the Fibonacci numbers and related topics".

I am a little uncomfortable with such obsessions. It is all too easy
to get hooked on numerology. Numbers have an insidious magic of
their own which tends sometimes to obscure the truth and to divert
proper investigation. But there is solace in the fact that this one, at
least, seems to be grounded in down-to-earth biology. If leaves did
not diverge, each would be shielded by the last one from the sun
and, with all its leaves on one side of the stem, plants would be in
danger of toppling over. The angle at which leaves do diverge is
probably not determined genetically – the only inherited instruction in any plant seems to be that they should diverge and the rest is
left to light, gravity and the growth rate of cells in the apical
meristem.

There are a host of other links between nature and number. The crystals of common salt are perfectly cubical, while snowflakes ring changes on the hexagon. The shell of the nautilus, a relative of the squid, has chambers arranged in a perfect logarithmic spiral. And the orbits of the planets round the sun are patterned on the shape of conic sections.

The fact that nature mirrors forms and relationships in theoretical mathematics, shouldn't come as a surprise. Even mathematicians are human and, like the rest of us, seem to have Earth's measure. We are all natural productions, imbued with an innate sensitivity to our planet's cycles and rhythms and, given the freedom to express this heritage, we tend to produce designs that reflect our origins. Though sometimes our busy brains need to be lulled into relative inactivity, forced to listen to the heart-beat, before we get the message.

In the middle of the nineteenth century, with Europe in the throes of various revolutions, the growing science of chemistry found itself in equivalent turmoil. John Dalton's contention that the elements differed from one another only in the mass of their atoms, was slowly becoming accepted and there were several rival calculations of relative atomic weights. Unfortunately, these differed so widely that even a simple compound like acetic acid, which only contains three elements, was being given nineteen different formulae by competing groups of chemists.

One of the first to try to bring some order to this confusion was the English chemist John Newlands, who suggested in 1864 that the relationship between the sixty-three elements then known was best defined by listing them in order of increasing atomic weight. But he was laughed at. The journals of his day refused to publish the paper, suggesting derisively that his arrangement of the elements by atomic weights made as much sense as listing them in alphabetical order.

It was not until five years later that the problem was eventually resolved, by a Russian and as a result, it seems, of a mid-afternoon dream.

Dmitri Mendeleyev, at the age of twenty-six, attended the First International Chemical Congress at Karlsruhe in 1860 and was impressed by discussion there of methods for calculating atomic weight. He became concerned with the problem of order amongst the elements and made a set of sixty-three cards listing the various chemical and physical properties of each, spending most of his spare time in attempts to arrange these in ways that made scientific sense. By March 1st, 1869, he had tried and rejected hundreds of possible patterns and that afternoon, following the failure of his latest system, he fell into an exhausted sleep – and woke up with the answer.

In order to make sense of the elements, Mendeleyev realised, all he had to do was arrange them, according to their atomic weights, in eight vertical columns. As soon as he did this, they fell into natural patterns, with elements of similar properties such as acidity, hardness and melting point, all grouped together. By evening of that momentous day, he had drawn up his celebrated Periodic Table in which the relationships between various elements were made clear by their regular spacing.

Mendeleyev refined and improved the table later, and in the ensuing century it has been extended and subdivided in a number of ways to allow for new discoveries. But despite the suggestion of over 700 other versions in the interim, his pattern remains the best and most useful yet devised. The strongest test of its validity has been the way in which gaps in the table, deliberately left there by Mendeleyev, have been filled by the discovery of new elements (we now have 107), whose existence the table clearly required. Mendeleyev himself pointed out three major holes, and accurately predicted the properties of gallium, scandium and germanium, which were not actually discovered until 1875, 1879 and 1885 respectively.

Mendeleyev's brainstorm clearly depicts an underlying pattern of nature, an order amply confirmed and explained by our new knowledge of the behaviour of electrons and protons. And it seems proper and satisfying that this insight, the basic shape of the "dream table", should have come to him in his sleep. It is not, of course, the only great inspiration to have been arrived at in this way.

In 1865, another mental breakthrough was made by the German chemist Kekulé von Stradonitz while travelling half-asleep on a bus. He was convinced that the atoms of chemicals, particularly the complex organic molecules, were arranged in special structural forms with their own particular properties. Starting in 1858, he successfully worked out the shape of a series of carbon compounds, but one in particular continued to elude him. He was unable to make sense of benzene, a substance discovered by Michael

28

Faraday in 1825 and in increasing use as a base for new synthetic dyes. Kekulé worried over the problem, but could find no solution until that day on the bus, when it seemed to him that he saw atoms whirling in a serpentine dance. As he watched the movement in his mind's eye, the tail of one long atomic chain was swallowed by its own head and took on the form of a spinning ring. This architectural vision gave him the clue he needed for description of a whole group of cyclic or ring compounds that still play a crucial role in organic chemistry.

There seems to be direct link between truly creative intelligence and the ability to dilute consciousness, to cut mental corners and practise unusual, lateral thinking in what amounts almost to a state of trance. All the most profound insights seem to flow from breaches in the barrier between waking thought, which tends to be conservative, and dream logic, which is essentially liberal. It cannot be purely accidental that Coleridge composed "Kubla Khan" in his sleep or that Mozart found his best musical inspiration rising like dreams, quite independent of his will.

It seems that, under conditions of dissociation, we have the chance to tune in directly to some of the world's basic rhythms, to become aware of the pattern behind the process. To know, in the words of Keats, that "what the imagination seizes as beauty, must be truth."[205]

Imagination

Most of what goes on in the physical world, is determined by a small number of universal constants, of things that do not vary. The value of these constants is now well-known, but it is only recently that we have begun to realise how many of them are the result of wildly improbable coincidences.

For instance, all natural phenomena are controlled, as far as we know, by just four fundamental forces – gravity, electromagnetism and two nuclear forces that have been called weak and strong. In recent years, valiant attempts have been made to describe the action of all four forces in a single unified theory – so far without conspicuous success. But the effort seems to be meaningful, because it is clear that the forces themselves cooperate to produce some astonishing coincidences.

Not least among these is the fact that the forces of gravity and electromagnetism are very delicately balanced. The bigger stars, those with a high mass, put out heat at a phenomenal rate, becoming in the end what are known as "blue giants". Smaller stars lose most of their more modest heat by the action of convection currents, sliding into a condition in which they are described as "red dwarfs". Both extremes are inherently unstable, but between the two is a very narrow range of star sizes that allow the sort of equilibrium provided by our far more kindly Sun.[76]

If the balance between the forces tipped in favour of gravity, *all* stars would be washed-out blue giants. If, on the other hand, it favoured electromagnetism, the universe would by now be studded with *nothing but* bright red dwarfs. And there would be no health in us.[402]

We owe our whole existence, and that of all typical stars it seems, to a wildly improbable numerical accident that equates two of the fundamental forces in the universe. And the coincidence doesn't end there. The number of stars in a typical galaxy is the same as the number of galaxies in the universe. The age of the universe, expressed in nuclear units, is the same as the number of charged particles in it. The electric force between two protons is the same as the gravitational force between them.[298] And all of these measurements, together with a long list of basic parameters, hover around the same unimaginably large number – the huge and mystic ten-to-the-power-of-forty (10^{40}).

Fifty years ago, the Nobel Prize-winning English physicist Paul Dirac said "such a coincidence, we may presume, is due to some deep connexion in Nature between cosmology and atomic theory." He and the astronomer Arthur Eddington were so impressed by the recurrence of this large and unlikely number that they built elaborate theories around it, ending however with little more than the simple and humble admission that "something strange is going on".[82]

Just how strange is only now becoming apparent.

Life as we know it is based on carbon, which did not exist in the early universe. In the beginning, there seems to have been nothing but a lot of hydrogen. Then came the big bang, which created temperatures high enough to make some heavier elements, but lasting only long enough, perhaps for just a few minutes, to produce large quantities of helium. The synthesis of heavy elements in any quantity had to wait until there were suitable stars of sufficient stability to cook the ingredients for the necessary billions of years. Which is precisely what happened and, as the stars reached the end of their natural lives, exhausting their nuclear fuel, they became supernovae, exploding violently and sending their varied contents spewing into interstellar space.

Amongst these elements, there seems to have been a surprising amount of carbon, formed as a result of yet another happy coincidence. Carbon nuclei come into being as a result of the rare and simultaneous triple collision of three separate helium nuclei. If just two collide in precisely the right way, they form an unstable nucleus of the hard white metal beryllium, which exists only for a very short period of time. And if a third helium nucleus strikes the temporary beryllium with exactly the right amount of energy, it too becomes incorporated – to produce carbon. None of this, however, can take place unless the resonance, the frequency of internal vibration of all three nuclei, is in complete harmony. But, as chance would have it, the thermal energy of a typical star, the temperature of its interior, lies at the one level that makes this not only possible, but inevitable.

And still the plot thickens. For the newly produced carbon to survive inside the star, it has to be prevented from combining even further or burning up to produce yet heavier elements such as oxygen. Fortunately for us, this prohibition is ensured by the further "accident" that the natural resonance of oxygen lies at a lower level than that provided by the combination of the first three helium nuclei. Events in the star conspire, somehow, not only to produce large quantities of carbon, but also to ensure that most of it stays that way long enough to be disseminated across the universe as one of the vital seeds of life. "Our bodies are formed," as Sir James Jeans once remarked, "from the ashes of long dead stars."[193]

Another scientific knight, the fearless Sir Fred Hoyle, has no doubts about the nature of all this cosmic coincidence. He calls it "a put-up job". In discussing the origins of life, he points out that neither carbon nor oxygen could ever have been produced in stars, unless their nuclear resonance was fixed at precisely the known levels. Nothing else will do. He concludes that "a commonsense interpretation of the facts suggests that a superintellect has monkeyed with physics, as well as chemistry and biology, and that there are no blind forces worth speaking about in nature."[170]

31

It is becoming difficult to deny that there is a hidden principle at work, organising the cosmos in a coherent way. And though the coincidences accumulate in fundamental physics and astronomy, these sciences seem to have no access to any reasonable explanation. Short of a resort to argument from design, to the assumption of the existence of an omniscient designer, our only chance of an answer seems to lie in biology, because the smallest tilt in the tuning of the cosmos would have made our kind of life and our existence impossible.

In many respects, our Earth is unexceptional. It may well be typical of a number of similar planets in place around similar stars. But the fact is that it is solid and situated near a steady Sun, when most of the universe is in a gaseous and highly unstable state. The chances of the conditions responsible for our hospitality coming together by accident, though small, are statistically inevitable. They had to happen somewhere. And having happened, acted it seems like a magnet for the seeds of life.

The laws of nature, in themselves, may have been sufficient to force such coincidence of the universe; but they do not, of themselves, necessarily tell the whole story. Things here fit so well that they seem to require at least an element of intelligent design. I cannot help feeling that there is something more to the process than blind chance, something that cries out for the exercise of what I can only call imagination.

Nature is restricted to the use of an alphabet which contains just 107 letters, one for each of the known chemical elements. In theory, this gives it sufficient complexity to say almost anything. In practice, most of these letters are inorganic curiosities that are hardly ever used. The whole living world is described in a shorthand alphabet of just four symbols, those which stand for carbon, hydrogen, nitrogen and oxygen. And an inevitable consequence of this restriction is that words and sentences constructed with such an alphabet have a tendency to be rather long; but they are, in addition, extraordinarily intelligent. The chemical passage that describes an elephant, for instance, not only contains all the necessary data for putting one together, but also assembles it so that it is able to react with the environment in a purposeful way.[117]

Einstein's awareness of this harmony in natural law led him to conclude that it "reveals an intelligence of such superiority that, compared with it, all the systematic thinking and acting of human beings is an utterly insignificant reflection."[91] James Jeans suggested that "the universe shows evidence of a designing or controlling power that has the tendency to think in a way which, for want of a better word, we describe as mathematical." It begins to look, he concluded memorably, "more like a great thought than a great machine".[193]

These are perhaps surprisingly religious statements from a pair of dedicated scientists, but neither was suggesting that the appearance of intelligence and design necessarily implies the existence of a designer. Each was celebrating, in his own way, a dawning realisation that the process of life itself is wonderfully cerebral.

We are one of the consequences of coincidence, the products of order picked out of the prevailing chaos by a cosmos that seems to exercise a bias in favour of such constructive accidents. And if we are intelligent, it may be because we are part of an imaginative process. Thoughts, even if they turn out to be only passing fancies, in the mind stuff of an intelligent universe.

Chapter Two

PERSON

Of all natural patterns, the one I think that moves me most, is the sight of a flock of wild geese.

A single goose passing high overhead carries with it a sense of freedom and adventure. "He is," in the words of Hal Borland, "the yearning and the dream, the search and the wonder, the unfettered foot and the wind's-will wing."[31] But a complete formation of geese is, for me, the epitome of wanderlust. Each one leaves me, no matter what I happen to be doing, wondering how long it will take me to pack my bags. And it's not just migratory restlessness, the knowledge that by dawn the flock will be in other climes. I don't feel the same way about swallows. There's something about the goose formation itself, that arrowhead symbol of limitless horizons, that hints at appropriate and meaningful adaptation. A sense not only of going somewhere, but of doing so together in the best possible way.

Observations of geese in passage, show that they invariably adopt a "vee" formation, flying on the same level, equally spaced out but not necessarily along arms of equal length. The important thing seems to be that the vee must have an apex – that the leading bird should always have others on either side.

It has been suggested that this characteristic formation is nothing more than a simple consequence of the fact that geese have immobile eyes on the sides of their head; and that, with the beak pointed forward, the best way to keep a neighbouring bird in full view is to take up a place just behind it, either to the left or right eye side. But direct measurement of flights of Canada geese shows that the angle between the arms of the vee formation varies even in a

34

single species between 28 and 44 degrees, which doesn't necessarily correspond with the fixed angle of clearest focus.[130]

Another theory suggests that the vee formation allows one goose, presumably a stronger and more experienced bird, to lead the way, cleaving a path through the air for the others. But, once again, field studies show that the leadership changes constantly and that this position, far from being reserved for wise old ganders, is in fact shared out amongst the younger and weaker members of the flock.[164]

The answer seems to be largely aerodynamic.

A recent computer study shows that there is an upwash beyond and behind the tip of a moving wing that can be useful to other birds nearby. If the spacing between wings is optimal, this saving in energy can be considerable. For instance, a formation of twenty-five birds can, just by adopting the most favourable formation, increase their effective range by 71 per cent.[229]

And this seems to be precisely what happens. Travelling geese usually fly in groups of around twenty individuals and invariably adopt a vee formation. If they flew in line abreast on a common front, the birds in the centre would enjoy twice as much uplift as the ones at the ends of the line. But as soon as the line is bent backwards, the ones at the rear begin to pick up additional upwash from all those in front, which effectively cancels out most of the disadvantage of their position. And as they travel, other small inequities which may exist are dealt with by regular and democratic changes of place.

The significance of this elegant aerial shuffle is that the geese automatically take the line of least resistance, letting themselves be shaped by natural forces. A bird that moves ahead of the vee line, will suddenly discover that it needs to exert more power just to keep up; and one that drifts out of the area of optimal uplift will find itself in an awkward patch of clear-air turbulence.

In addition to the pattern of formation, the computer study also predicts that maximum efficiency for a group of birds can be achieved by a cruising speed 24 per cent lower than that which best suits a solo bird. And observation indicates that geese have not only discovered this to be true, but at slow speed are able also to take better advantage of the existence of any tail winds.

The easy and the proper thing to do, is also the one that represents the most meaningful adaptation to the prevailing conditions. It is the move that makes environmental sense and emphasises the many benefits of being social – of becoming part of a larger and more adaptive society.

Computers help us to understand the applied mathematics of social advantage, but there are other aspects of getting-together-for-the-greater-good that are not as easily analysed. Something happens to groups of individual plants or animals that makes them more than the quantitative sum of their separate parts. They undergo qualitative changes that can seldom be predicted from a knowledge only of the nature of the component parts. They go through an almost magical transition that can produce some extraordinary results.

Cells

Life needs energy. And, in the final analysis, the energy on which all life on Earth depends, is provided by the Sun. Every course in biology begins with the phenomenon of photosynthesis – our process for tapping the energy of the Sun – but few bother to remark on the extraordinary coincidences that make it possible.

Earth is bathed in vibrations, most of them coming from the nearest star, our Sun, which is involved in a thermonuclear blaze that destroys four million tons of matter every second. Radiations from this conflagration range from very long frequency radio waves, some a mile wide, down to gamma rays so short that sixteen million million fit into an inch. Most of this broad spectrum just happens, if you still believe in blind chance, to be filtered out by our atmosphere. The gases in Earth's security blanket mop up radiation like an electromagnetic sponge, leaving a narrow window that permits free passage only to waves whose length lies roughly between 0.0004 and 0.00004 of an inch – the ones we call visible light.

It also "just happens" that wavelengths longer than this, those known as infrared, are hard on sensitive membranes. While shorter ones, those in the ultraviolet range, carry so much energy they unzip the long chain molecules of most proteins, making life altogether impossible.

The atmosphere begins to look more and more like an artefact – something created and maintained for the express benefit of living things. Something which has, through almost 4000 million years, contrived to keep the temperature of earth's surface within astonishingly narrow limits – which, once again, just happen to be precisely those that life requires.

By further coincidence, living things have evolved pigments which resonate only to wavelengths which lie in the narrow band of visible light that does pass unhindered through our planetary defences. A circumstance which George Wald, the American chemist who won his Nobel Prize for work on the nature of vision, says must be true throughout the universe. There cannot be a planet, he maintains, on which either photosynthesis or vision, or anything like them, occurs in the infrared or the ultraviolet wavelengths, or in any other wavelength, "because these radiations are not appropriate to perform such functions".[387]

Once again, it seems, evolution has managed to get things just right.

Photosynthesis itself is a remarkable process, one which depends on substances far more complex than any of the raw materials with which it works. At the heart of these is chlorophyll, a green pigment which makes it possible for plants to build up carbohydrates from carbon dioxide and water. This pigment is supplemented by red and brown variants better able to deal with the even more restricted wavelengths which filter through to life underwater, but all have immensely complex atomic structures based on a ring of four chemical elements around an atom of magnesium. And all rely, for their formation, on the action of two distinct chemical systems, each with its own peculiar enzyme. The presence of one without the other would have been useless. Everything is contingent upon the elaborate coincidence of both enzyme systems being precisely where they would be needed. And it becomes difficult to believe that anything so intricate and so appropriate could have appeared so early on in the history of life on Earth, unless there was some innate predisposition for it to do so. "It seems," in the words of Gordon Rattray Taylor, "that in the biological sphere at least, there is a built-in tendency to self-assembly at the most elementary level."[368]

There is widespread evidence of the ability of organisms to assemble and re-assemble themselves under the most difficult circumstances. Marine sponges are an excellent example. They consist of several different types of cells that gather around supporting skeletons of superb geodesic design, working together so closely and in such consistent forms that these aggregations are classified as distinct individuals and given recognised specific names. Any such consortium can be taken from its base on the ocean floor, cut up into convenient portions and squeezed through silk cloth with a weave fine enough to separate every cell from its neighbour. And yet, if this amorphous gruel is allowed to stand quietly for a while, it soon gets together and reorganises itself – and before long the complete sponge, one impossible to distinguish from the original, reappears like a phoenix and goes straight back into cooperative business once again.

In one fascinating experiment, a red encrusting sponge that is classified as *Microciona prolifera*, and the yellow sponge *Cliona celata*, were sieved together and their cells thoroughly mixed. But after just twenty-four hours, the red and the yellow cells had successfully separated, reorganised and reassembled themselves, getting back together in their original forms. Quite apart from this impressive piece of biological management, the experiment raises awesome questions about identity. Two distinct living organisms existed at the beginning, but who was who, and was either alive or dead in the blended soup? The cells were clearly living, but where were the organisms? And what conclusions can be drawn from the strange fact that a few red cells turned up, apparently quite happily, built into the new yellow sponge?[161]

Sponges are marginal things, lurking on the borderlines between· plants and animals, between colonies and actual creatures. But the same kind of organisation is evident in more complex organisms. Sea urchins, the spiny bane of barefoot bathers, are elaborately organised. They have well-developed digestive tracts, tube feet for getting about, vascular systems for the transport of nutrients, and a ring of thorny plates arranged into an intricate skeletal scaffold. The variety of styles in such armour is dazzling. Living and fossil forms of urchin have adopted every imaginable pattern, like fashion models trying on dress after dress. There seems little reason for many of these experiments, few have any direct benefits to offer in the battle against extinction. Far from being involved in a dour Darwinian struggle for survival, their evolution and radiation look, in the words of one naturalist, "more like a glorious romp".[368]

And yet, each of these variants seems to have a clear sense of identity. If kept in an artificial medium that lacks the calcium necessary for constructing their elaborate skeletons, all urchins collapse into a helpless stew of unconnected cells. But they nevertheless, somehow and somewhere in the chaos, retain a memory of past and proper form. As soon as the essential calcium is reintroduced, the cells regroup and reassemble themselves into the familiar armour-plated globe of a typical sea urchin.

This ability to instil order is the most vital and peculiar characteristic of living things.

In the non-living world, atoms assemble themselves into molecules, but with the exception of crystals, there is no evident tendency for the molecules to go on and arrange themselves in turn into more elaborate structures. Living material, on the other hand, shows what amounts almost to a compulsion to do so.

At the University of Pennsylvania, Stewart Kauffman has produced a computer simulation of the evolution of cells from genetic instructions. Assuming the existence of ten thousand genes, which is a conservative estimate for most cells, he calculates an unimaginably vast number of possible combinations – something approaching 10^{3000}. But as soon as restrictions are introduced, based on the known ways in which cells are able to interact with each other, this unwieldy diversity settles down very quickly into recognisable patterns. The programme spontaneously confines itself to about a hundred different combinations – which is roughly the same as the number of different cell types actually found in complex organisms.[121]

There do seem to be built-in constraints which underlie the process of natural selection, channelling it into a relatively small number of relevant forms. A flat sheet of growing cells, for instance, is virtually compelled, when those cells in the centre increase in size more rapidly than the rest, to produce a bulge. And this bulge can only grow in like a pouch, or stick out like a thumb. And once a tissue passes any such point of decision, it is committed to a certain line of development and cannot turn back. Any such system, as it becomes more complex, necessarily becomes more orderly, more restricted in its options, less likely to go off at an unpredictable tangent. It is compelled towards certain natural ends, which in effect design themselves.[89]

The astronomer Eric Jantsch suggests that things are like this everywhere. That we have stars and quasars and black holes simply because the properties of matter make their production inevitable.

This is, he says, partly a result of environmental constraints, a natural consequence of the laws of physics. But partly also due to a built-in necessity for things to grow in a certain way in what he calls this "self-organising universe".[192]

So the cells in the growing tip of a plant fall into patterns dictated partly by the instructions of the genes, which decide on the shape and colour of leaves appropriate to that species. But the arrangement of the leaves, the angles at which they diverge from one another, take a natural and inevitable and orderly pattern of the kind defined by Leonardo Fibonacci.

The only rule governing the combination of instruction and inevitability seems to be the tendency of all things to move towards equilibrium. The action of this principle, which cannot be called a force, because it is impossible to measure or quantify, is best expressed by Arthur Schopenhauer in a memorably prickly image:

On a cold winter's day, a group of porcupines huddled together to stay warm and keep from freezing. But soon they felt each other's quills and moved apart. When the need for warmth brought them close together again, their quills again forced them apart. They were driven back and forth at the mercy of their discomforts until they found the distance from each other that provided both a maximum of warmth and a minimum of pain.[331]

In human terms, loneliness brings people together into societies, where their faults drive them apart again, until they eventually find an optimal individual distance that is ritualised in elaborate codes of customs and manners. And so it seems to be with the basic building blocks of all living organisms. Cells seek the proximity of other cells, up to a point, finding arrangements which allow diversity of function while protecting their separate identities.

The net result is a range of combinations which is severely circumscribed, but sufficient also to produce amazing new properties and talents.

Organs

The German biologist Ernst Haeckel was one of the last great natural philosophers. He was the first to use the term "ecology", including in it a visionary view of the forces which hold simple cells together in complex functional wholes, which he called "persons".

A person, in this sense, is an aggregate. A group of cells, not necessarily of the same kind, which work together towards a common end. We have no record of the number of such associations which must have formed during the course of evolution and disappeared without trace, but one at least of these early colonies proved so successful that it can still be seen rolling smoothly along through the surface layers of most freshwater pools.

Volvox is about the size of a pinhead, a determined little green person consisting of around two thousand cells arranged into a slightly lemon-shaped sphere. Each of the cells has two long whip-like flagellae, relics of a time when they swam off on their own in search of the sunlight on which they feed. In fact, if any one of these cells is artificially detached from the colony, it still swims away, apparently quite content to do so again. But as colonists, the cells have become planted, tails out, around the surface of a sphere, taking on the hexagonal shape which, in everything from plant pith to honeycomb, has turned out to be the most efficient form of biological packaging. Each cell is separated from its six nearest neighbours by a matrix of insulating jelly, but is nevertheless connected to them on all sides by fine protoplasmic strands that form a continuous network over the entire surface of the colony. The centre of the ball of cells is hollow.

Perhaps the most interesting feature of a *Volvox* colony, the first sign of its budding "personality", is the fact that it has a definite polarity, a head-and-tailness. As the flagellae with which it bristles begin to beat, the tiny globe rotates about its axis, spinning just as earth does, keeping one side up. The cells in this northern hemisphere, which are larger and greener than the others, are kept permanently facing the sun by coordinated beating of the colonial whips. These send the sphere spinning in one direction for a while and then stop it dead in its tracks, or even reverse the motion, fine-tuning their joint response to keep the collection in the light.

The cells in the southern hemisphere have other talents. From time to time, one or more of these begin to go through a series of repeated divisions. The result is a pouch which bulges inward into the hollow centre of the sphere, growing until it is large enough to become completely detached, floating free as a new and distinct daughter colony. And the process sometimes repeats itself again and again, producing several generations of *Volvox*, clearly visible one within another like a nest of Russian dolls. Each of these daughters has an independent destiny of its own to fulfil, but it is one that cannot be realised until it has first solved an awkward problem.

In the process of invagination, as the pouch extends into the interior of the mother colony, the constituent cells of each daughter end up with their whips on the inside, where they are clearly useless for any kind of locomotion. The new colonies, however, resolve their dilemma in a most revealing way. One side of each caves in, like a rubber ball on which you press your thumb, until it meets the far side. Then a hole appears, through which the young colony pours itself like a sock being turned inside out, passing in the process through a fascinating tube-shape – one that anticipates the basic body form of far more complex creatures. Showing, in one highly suggestive manoeuvre, that this primitive little plant, it is after all only an alga, already contains the sort of genetic instructions necessary to lead to most animal life, up to and including our own.[29]

It is clear, right from this early stage of evolution, that the process is to a large extent self-governing, shaped in ways that eerily predict its future pattern and direction. Haeckel, in his pursuit of the person, noticed that the embryos of vertebrates seemed in their individual development to review all the stages of their collective history. "Ontogeny recapitulates phylogeny," he said. And there is some truth in this. A developing embryo does seem to shuffle through the cards of evolution, looking at a variety of patterns in sequence, selecting the bits that suit it best. But it would probably be more accurate and more productive to invert Haeckel's maxim and suggest instead that "ontogeny anticipates phylogeny". The seeds of evolution were sown long ago – and much that has happened since, has been almost inevitable.

In theory, all cells of a kind are created equal. But in practice, some are more equal than others. The beginnings of specialisation evident amongst the flagellated cells of *Volvox* is taken a fascinating step further amongst colonial amoebae.

Generally speaking, amoebae are independent and solitary, pushing out parts of their substance in an apparently aimless fashion as they ooze about damp places. There are some species, however, that can take communal action, turning to social living when food is short. *Dictyostelium discoideum*, better known as a slime mould, normally creeps through liquid films on the forest floor, engulfing bacteria and dividing relentlessly again and again, every three or four hours. But when threatened with starvation, a randomly speckled field of such individual amoebae, all more or less equally spaced out, can turn in minutes into a feathery pattern as the cells begin to stream in towards a central point, where they form a sausage-shaped communal slug delightfully known as a grex.

There are forty thousand separate amoeboid cells in such a colony and yet the collection is capable of behaving as though it were a single organism, gliding off at the great rate of one millimetre an hour in search of a warm bright place in which to breed. There is differentiation within a grex, with cells at the front end taking on a distinct identity, "knowing" somehow that it is their task to lead. John Bonner at Harvard University stained one such slug with a harmless dye and grafted its front end on to the back of an unstained slug. Within minutes, the stained "head" cells moved rapidly up the mixed slug, sweeping through its fabric like a band of colour until they were able once again to reach their proper place in the lead.[30]

When a determined grex eventually comes to rest in the spot of its choice, it performs an even more extraordinary feat. The colony sorts itself out into working parties, each with a separate function, and seals a select group of its fellows into a capsule that is hoisted up into the air at the end of a long, thin, well-anchored stalk. A group of identical and originally isolated cells succeed in the complex tasks of getting together, deciding on a course of action, undergoing differentiation and a well-orchestrated division of labour, dedicating their joint efforts in an extraordinarily altruistic way to the promotion of survival in a chosen few of their number who become, in effect, the occupants of a space capsule designed to travel to a more auspicious habitat. The original cells, between them, produce an entirely new unit designed to carry out this specific function. In other words, they get organised – they make an organ.

An organ in biology is defined as a number of cells gathered into a structural and functional unit as part of a living system. Leaves and kidneys are advanced examples of such units. We understand a great deal about how they work, but still know very little of their origins. For them to exist at all, it was necessary for evolution to

43

make a quantum leap from being simply colonial to becoming truly metazoan. The term means "after animals" and refers to all creatures whose bodies consist of many cells, up to and including humans. It reflects, in its implied exclusion of multicellular plants, an ancient bias, a deep-seated unwillingness even amongst biologists to recognise vegetables as behavioural organisms. But the problems are as real for seaweed as they are for seagulls – both have had to bridge the wide gap between simply getting together and really working together as functional wholes.

The turning point for all such colonies of cells came when their mass grew so large that some cells were cut off from the outside world. At first, these simply died, as bacteria in the centre of rapidly growing colonies still do. But then something mysterious and important happened. The cells on the outside of the bundle began to transport food and raw materials to those inside, responding in new and sensitive ways to signals, to distress calls, from within.

The best understanding of this momentous development comes perhaps from work on tissue culture – the delicate art of getting small groups of cells to live and multiply in isolation. At Cambridge University they have succeeded in persuading cells from an extraordinary variety of sources to take nourishment and grow in a single drop of liquid on a microscope slide. A tiny fragment of minced chicken heart under these circumstances soon sorts itself out into four distinct cell types. The first and most numerous of these is a structural cell capable of forming bone and muscle. The second is a latent skin cell that tends to join up with its fellows edge-to-edge, forming a pavement or covering. The third are sensitive cells that stretch into incipient nerve fibres, and the fourth are mobile cells that wander around like free-living amoebae.

The Cambridge study shows that this subdivision into the same four basic cell types takes place regardless of the origin of the tissue sample. Chicken heart, rat tongue, eye of newt and toe of frog – in isolation all tend to break down in the same way. They lose their complex identities, slide into anonymity and then start their own little evolutionary sequence, differentiating into four apparently historic forms – each one the precursor of an essential metazoan organ.[415]

The forces responsible for this division of responsibility remain mysterious, but the implications are clear. All complex creatures

44

consist of a strangely limited number of cell types and are predisposed to develop along certain clearly defined lines. They have an organic tendency to form structural units that work in a particular way – and the shape and nature of this "organisation" is largely predictable.

Life knows its lines and needs little prompting. And reading between the lines, there is evidence of a surprisingly strong, widespread and stable personality. Something we all share.

Organisms

Traditional biology looks at multicellular plants and animals as complex pieces of machinery. It describes these as "organisms" and highlights the nerves and vascular tissues responsible for controlling such organisation. But, at the same time, it also severely limits the degree of response believed to be possible for each kind of complexity.

The tendency has been to reserve what have been called "higher functions" for the more advanced forms – meaning man and a handful of favoured mammals. In recent years, however, this warm-blooded, big-brained élitism has been challenged by demonstrations of what appears to be elaborate sensitivity in some very humble species. And every dog-fish and dog-fennel has had to be given its day.

The turning point came in 1966 with the activities of an ex-intelligence agent in New York. It was in that year that Cleve Backster started his experiments with plants and polygraphs, connecting various vegetables up to sensitive electrical equipment so that they produced what appeared to be appropriate responses to specific stimuli.[5]

The story of his work is now part of the folklore of the paranormal, but remains a point of scientific contention. Subsequent studies, most notably a carefully controlled series of tests at Cornell University, have failed to find corroborating evidence of the "primary perception" that Backster claimed for all living things – and most biologists have been able to dismiss the whole phenomenon as a meaningless aberration.[169] The sighs of relief at this deliverance have been almost audible, but there remain a few of us who, having taken part in a broad range of experimental work with plants, find it difficult to ignore all the results quite so easily.[399]

45

There is no question that much of the work that seems to demonstrate awareness in plants, has been sloppy. Artefacts introduced by the equipment and by changes in temperature and humidity, abound. But there are coincidences of stimulus and response that remain intriguing and make it impossible just to discard all the data as pseudoscientific nonsense. I suspect that there is something going on that is still well worth pursuing and I offer a detailed analysis of one particular piece of natural history that I find most intriguing.

It comes from Africa, where a long series of drought years during the early 1980s had a devastating effect on local fauna and flora. Populations everywhere tumbled, but wildlife management officers in South Africa were nevertheless somewhat surprised to learn from farmers in the north-western Transvaal that large numbers of kudu were dying. These majestic, spiral-horned antelope are normally amongst the more drought-resistant species. Despite their size, greater kudu *Tragelaphus strepsiceros* flourish even on the desert margins, succeeding somehow in browsing a living off vegetation too sparse to support most domestic livestock. So the scientists went to take a closer look.

They found that the deaths were taking place in thornbush country which was indeed very dry, but by no means poor enough to allow hardy kudu to die of starvation. Post-mortem examination of several dead antelope showed that their internal parasite loads were normal and none was suffering from any recognisable disease. Many had apparently adequate quantities of leaves in their digestive tracts, but even these ones were emaciated and seemed to have died of malnutrition. So the problem was passed on to Wouter van Hoven, a physiologist at the University of Pretoria.[383]

Professor van Hoven started by analysing the nutritional value of the kudu's usual foods. He found that the protein content of the leaves involved was high, averaging around 14 per cent – enough to keep even the largest antelope healthy. But then he discovered that the protein content of the kudu's droppings was equally high, indicating that the food had passed straight through their intestines, almost undigested.

Under normal circumstances, leaves are processed in the rumen or main stomach of a kudu, where microbes secrete enzymes which organise a type of fermentation. But nothing like this was taking place and, instead of producing new and useful fatty acids, the kudu were burning their own body fats, locked into a destructive cycle that led inevitably to their own deaths.

The search turned back to the food, to a closer look at the leaves from a dozen species of common bushveld trees – and it soon became apparent that, in addition to their high protein levels, all the plants being eaten were loaded with a group of chemical compounds known collectively as "tannins". The name comes from the process of tanning, in which fresh hides are turned into more durable leather as their protein molecules are persuaded to combine with certain chemicals to form an insoluble substance that cannot be destroyed by micro-organisms. And the presence of precisely these chemicals in the diet of the kudu was effectively "tanning" their insides, turning off the microbes that encourage normal digestion.

There is, in natural history, a long-standing relationship between plants and the herbivores that eat them.[161] As evolving animals develop new ways of exploiting vegetable foods, the plants have succeeded in keeping beastly appetites within reasonable bounds by inventing a variety of defensive mechanisms. The most obvious of these are physical deterrents such as heavy bark or sharp thorns, but there are also an impressive array of chemical constraints, including the tannins. These not only make the plants bitter and unpleasant to the taste, but have been carefully designed to interfere with a herbivore's well-being. Some tamper directly with a diner's digestion, producing after-effects which can be lethal. Others have a delayed response, one passed on to the second and third generations, particularly of herbivores such as caterpillars which have been found, after eating tannin-loaded leaves, to produce fewer egg masses and weaker, shorter-lived offspring.[319]

Production of such chemical weapons is very effective, but it is an expensive strategy for the plants involved. A few species, notably the Australian eucalypts, which must have been under almost constant attack, have such a high and permanent concentration of condensed tannin that it was customary, before insecticides, to use their leafy twigs to keep flies away from colonial homes. But plants do not use these tannins for anything else but deterrence and most cannot as a rule afford to go on churning out such complex substances on the off-chance that the appropriate pest or predator

47

might happen to pass by and drop in for a bite. So what a number of species have done is to develop ways of producing their weapons only when the need arises.

Studies on the west coast of North America have shown how this process works. David Rhoades of the University of Washington selected pairs of red alder *Alnus rubra* and sitka willow *Salix sitchensis* growing in natural forest stands. One member of each pair was chosen at random as a control tree and the other was deliberately loaded with egg masses from the tent caterpillar *Malacosoma californicum*. A few weeks later, the test trees were swarming with voracious larvae.[303]

At first the insects flourished, but before long it became apparent that the colonies were growing more slowly than usual, constructing very small sleeping tents and doing surprisingly little damage to the trees. Analysis of the leaves showed why. Both alder and willow had, over a period of weeks following the initial attack, dramatically increased the amount of a chemical called proanthocyanidin in their leaves – and this was slowing the caterpillars down. Rhoades demonstrated the direct effect of the substance by raising similar insects in his laboratory and feeding them either on leaves from the infested trees or leaves from control trees. In every case, caterpillars eating leaves from trees that had already been attacked, and been given time to produce their chemical response, grew more slowly and became more susceptible to disease. The larvae in the test trees were prevented by artificial barriers (known as "tanglefoot" bands) from migrating to other trees nearby, and by mid-summer every single one of the insects was dead – without damaging more than 12 per cent of the leaf area of their vigilant hosts.

In Africa there was an equivalent situation. The shy kudu normally browse over wide areas, nibbling a bit here and there as they melt away through the bush. Rocks, rivers and barbed wire mean nothing to an antelope that can leap six foot fences without visible effort. But during the last decade an increasing number of game-conscious farmers have been erecting more formidable barriers to keep their local wildlife from straying into areas where they run the risk of being shot. The motives are admirable, but the effects have been unfortunate. Confined to limited areas, like caterpillars stuck in a single tree (the critical density seems to be around 100 acres per kudu), the antelope are forced in times of drought to feed more often than they would like from the same few trees.

The trees don't like it either. Observation of kudu in an undisturbed area shows that they browse on any one bush or tree for less than two minutes, moving on quickly to the next even though ample and apparently equally appetising leaves remain within easy reach on the first. Why not continue eating at each feeding site until all the available leaves have gone? The answer is simple. The trees won't allow it. They put up their chemical defences – and they do so with astonishing speed.

Kudu are substantial creatures, standing five foot high at the shoulder and weighing over a thousand pounds. They are not gentle feeders, stripping leaves and bark from branches, often breaking off twigs in the process. Something a plant is bound to notice. Wouter van Hoven wondered if he could imitate the process and took his students out on tree-beating expeditions. The first step was to tiptoe up on a tree and gently take a sample of its leaves. Then, acting on his instructions, a mob of students moved in with an assortment of belts, whips and canes and thrashed the hapless plants, damaging and stripping their leaves. And finally the scientists stepped in again and took further samples of the leaves at fixed intervals.

The results were dramatic. All the abused trees produced extra tannins and moved these into their leaves within minutes. After fifteen minutes, the concentration of tannin in weeping wattle *Peltophorum africanum* had risen by 44 per cent; in mountain sumac *Rhus leptodictya* the increase was 76 per cent; and in the hookthorn *Acacia caffra* it was almost doubled to 94 per cent. An hour after being beaten, the defensive chemical content of wattle leaves had risen by 256 per cent, while that in the acacias soared to 282 per cent above its initial level. Wild pear *Dombeya rotundifolia*, bushwillow *Combretum hereroense* and guarri *Euclea undulata* all showed similar responses and it took at least 24 hours, in some cases as much as 100 hours, for any of the plants to relax and return to a state of tasty equilibrium.[303]

Under normal circumstances, the rush of tannins to the leaves becomes appreciable to the kudu as a change in taste within the first few minutes of beginning to browse, and it moves on. But when high fences restrict the antelope's movement and drought restricts its choice, it is forced to return to a limited range of feeding sites at shorter and shorter intervals. Browsing pressures intensify, tannin concentrations remain dangerously high, digestion suffers and animals starve and die. It is significant that the Pretoria team found a number of dead kudu actually hanging from the offending fences, impaled as they made final frantic attempts to move on to fresh fields with access to more palatable and less paranoid plants.

This research begins, for the first time, to make sense of some bits of agricultural folklore. Farmers in Britain and France, where the walnut is held in high esteem (the generic name *Juglans* is derived from *Jovis glans* or "Jupiter's nuts"), have always believed that the trees bear heavier crops of fruit if the trunks and branches are attacked every now and then with a stout walking stick. The saying actually makes the scurrilous claim that "both women and walnut trees are best for an occasional beating," but it begins to seem possible that the trees at least might respond to such chauvinism by chemical changes that persuade them to set more seed in the hope of surviving in some future form. It could be significant that walnut leaves already contain a chemical that renders them almost immune to attack by insect pests.

And I am impressed by the fact that all of the garden and nursery enthusiasts I know – the ones with really "green fingers" for whom plants seem prepared to do almost anything – treat their charges with an almost avuncular brusqueness, clipping, pruning and giving their leafy crowns a rough and cheerful tousle every time they pass. Whereas my hesitant and reverential flower-bedside manner seems to lead to nothing but rootrot and leafblight and terminal decay.

Having solved the riddle of the dying kudu, Wouter van Hoven found himself faced with an extra dilemma. He was careful, in the tests on plants that were deliberately maltreated, to take additional samples from other similar specimens nearby as controls. And he found, to his astonishment, that if these undamaged trees were anywhere near the injured ones, they showed a sympathetic increase in tannin concentration. A small hookthorn of the same size as the one that was thrashed and which stood over six feet away, showed a 42 per cent increase in tannin after three hours. A silver oak ten feet away, produced 14 per cent more tannin within an hour. Even the "weeping" of wattle was registered by two other trees within ten feet, which produced high concentrations of tannin soon after a student had belaboured their neighbour with his belt.

50

Van Hoven was unable to explain this apparent ability of trees to raise the alarm and alert others of their kind to the danger of imminent attack. It is anyway just the sort of subject likely to cause equivalent alarm amongst editors of the more conservative botanical journals. So he simply published his observations under the title of "Tree's secret warning system . . ." in a popular magazine produced by the South African National Parks Board, where it seems to have gone largely unnoticed.[383] But it need not languish there, because there are comparable conclusions in footnotes to the Washington study on alders and willows – and in another piece of recent research on polar and sugar maple trees in the woods of New Hampshire.

David Rhoades in Washington took care to test the leaves of his trees by feeding them to healthy caterpillars under laboratory conditions. He found, as we now know, that leaves from trees under attack were bad for the insects. But he also discovered that leaves from control trees, still unattacked by pest or man, produced equally poor results unless these samples came from a good distance away. He was unable to find any evidence of root contact between the infested and uninfested trees and concluded, with the addition of an exclamation mark (a point of punctuation rare in learned papers), that "the results may be due to airborne pheromonal substances!"[303]

The suggestion that plants may communicate with the aid of hormones that drift through the air is nothing short of revolutionary. It makes them look much more sensitive, far more like animals, than the textbooks would have us believe. If trees can indeed warn each other of danger and respond to such warnings in appropriate and meaningful ways – there is little to separate them biologically from the vervet monkeys that use and react to social alarm calls at the sight of eagles, leopards and snakes. The only thing that prevents us from being reasonable and honest, and classifying alders and willows as truly social creatures, is the fact that they are incapable of running away. But given their relative immobility, it becomes increasingly difficult to deny that the actual response of the plants in question is highly appropriate and very impressive.

This is one of the reasons why I find it hard to be as sceptical about work on plant awareness and "primary perception" as I perhaps ought to be. It is true that none of the work, even the studies I have outlined here, can be regarded as unequivocal and providing final and positive proof. But they are highly suggestive and ought to lead to a careful re-appraisal of some hoary old prejudices – particularly the one which draws a hard and fast line between the plant and animal kingdoms and completely denies the former access to any of the talents and abilities we reserve for the so-called "higher" species.

There is good reason to presume, at least as a working assumption, that some kind of awareness is part of the experience of all living things.

Identity

During the course of evolution, plants and those that eat them have grown together, keeping a delicate balance in the matter of advantage. Each makes continual adjustments, keeping pace as far as possible with innovations introduced by the other.

The evidence shows that production of defensive chemical compounds by plants tends to go hand in hand with the development of passive avoidance mechanisms, or of active detoxification techniques, on the part of the herbivores.[161] Kudu simply move on to new pastures, but less mobile creatures such as the marsupial koala have had to come up with direct answers to the extremely high tannin content of the eucalypt leaves on which they feed. The koala solution remains a mystery to us; but it, or something like it, must be known to the host of insect predators which consume up to 40 per cent of the annual leaf production of over 300 species of Eucalyptus in Australia. This sounds one-sided, but it has to be said that the plants have succeeded in finding a way of making even this formidable onslaught work at least partly to their advantage.

Many trees fight for living space, for their fair share of light and nutrients, by producing substances called allelochemicals which inhibit the growth of rivals. These botanical boundary-marks are normally released from the leaves as volatile substances which settle on trespassers that come too close, or exude from the roots and percolate menacingly through the surrounding soil. In desert sands it is common to find plants so evenly spaced out by mutual chemical inhibition, that they seem almost to have been arranged by hand. Eucalypts in the drier parts of Australia, without expending any more energy than absolutely necessary, have solved their

52

territorial problems by letting insect pests do most of their legwork for them. The insects may have found ways of dealing with the harmful chemistry of plant deterrents, but lurking in amongst the potentially toxic tannins are allelochemicals which pass unaltered through the digestive process, and rain down all around the host tree as a major component of insect droppings or "frass". This noxious carpet, which can accumulate at the rate of a ton per acre per year, ensures that the area beneath many eucalypts is bare and free of all potential competitors.[345]

This nice little bit of one-upmanship appeals to botanists, but it is more than a biological curiosity. It highlights a major mystery – which is how trees ever survive at all.

The problem is that most trees live a long time, some of them (like the bristlecone *Pinus aristata*) for over two thousand years. And throughout this time they are open to attack and infestation by insect predators that breed at least once during every one of those many years. In theory – mutation, variation and natural selection, which act on each new generation in turn, ought to give the pests an unbeatable advantage, enabling them to overcome any defence with which the tree may have begun its long life. By rights there ought not to be any trees left anywhere, let alone ones that go on for millennia defying logic and the worst that rapid and adaptive invertebrate evolution can hurl against them.

Conventional wisdom in biology holds that somatic cells and reproductive cells have separate destinies, and that only changes which take place in specialised cells of the germ line can be passed on to future generations. Bodily changes that occur during the life of an individual, no matter how useful or advantageous these might be, cannot be perpetuated. The inheritance of acquired characters is impossible. But is it?

The tip of each growing shoot in a plant contains a mass of cells called the meristem which divide constantly to form the tissues of stem, bark, leaves and thorns. These make up the body of the plant. But the same meristem also produces flowers with ovules and pollen, the germ cells of that plant. There is no clear distinction between somatic and reproductive tissues. And no reason why all the growing shoots in a large plant like a tree, which are subjected to a variety of influences, should remain genetically identical.[55] The evidence suggests that they do not.

All the pink grapefruit in the world come from one branch on an

53

otherwise normal tree that suddenly produced this attractive variation. Seedless grapes and navel oranges are similar products of random changes which took place on traditional trees and were singled out for propagation by observant horticulturalists. Such alterations occur and are perpetuated equally well in nature.

Doug Gill of the University of Maryland has been studying witch hazel *Hamamelis virginica* and its ability to deal with seasonal attack by aphids. He has discovered that trees differ only slightly in their degree of infestation, but that there are major, and statistically significant, differences between the different branches on a single tree. Branches that are heavily infested one year, tend to be equally heavily attacked the following year – while branches that seem to have some defence against aphids, carry this resistance successfully through the winter and pass it on, through their seeds, to new generations of tree.

Trees, it seems, may be more complex than we give them credit for being. A large one may carry 100,000 meristems and the chances are that, given even ordinary rates of mutation, ten of these will undergo genetic change. Which means that, at any one time, up to ten branches in a great oak could each carry a distinct set of genetic instructions and look and behave differently. A fact which gives such a giant back some of the initiative it needs to keep pace during its long life with inventive, short-lived and rapidly multiplying pests.

The evidence suggests that big trees are actually genetic mosaics in which each shoot has the potential of a new generation – cutting evolution free from the classic constraints of normal generation time. Trees live as long as they do, simply because they keep on adapting and changing. They may be stuck in a hole in the ground, apparently at the mercy of all that moves, but they can be extremely resourceful.

It has been argued that millions of dandelion plants, which have reproduced vegetatively to cover many square miles with the products of a single clone, in fact represent just one genetic individual. A giant organism with a fixed genetic composition that just happens to be spread out rather thinly over a wide area. Compared to it, a great tree with its genetic mosaic ought to be considered, not so much as an individual, but as a complex society – an organisation rather than an organism.

A tree can be more than a tree. It can be a collection of variations on a central theme such as oak or poplar. Which begins to make sense of another botanical mystery.

Many trees produce all their flowers simultaneously. They burst into crimson flame like the poinciana or are totally enveloped in cool blue blossom like the jacaranda. It is usually assumed that this mass display is necessary to attract insect pollinators, who might have difficulty, without such a beacon, in finding a flowering tree in time. In tropical rain forest, where each jacaranda may be separated by half a mile and millions of other trees, from the next one, this makes some sense. But studies on bees and other insects attracted to such trees, show that they bring in very little if any foreign pollen.

With such a mass of feeding sites all in the same place, insects spend all their time in one flowering tree and pollen is simply moved around within the tree itself. If all the branches and flowers of such a tree are genetically identical, then this process is a waste of time. It is equivalent to not mating at all. But if the tree is a genetic mosaic, if it is in effect a community of slightly different individuals, the transfer of pollen will be useful and productive, leading to still more variation as a result of new combinations.[55]

Suspicion grows that earlier studies, which found most trees to be self-incompatible, may not have looked far enough afield for their samples of pollen. Flowers on the same branch are likely to share the same genotype and therefore fail to set seed. But pollen taken from flowers on another branch on the far side of a large tree, is likely to be more productive.

It is likely too that we have been equally short-sighted about the degree of selection and control that plants can exercise in mating. Many more pollen grains alight on a flower stigma than are required to fertilise the ovules lying below it, and evidence is accumulating to show that plants can and do discriminate between rival grains. They ought not to be seen simply as passive receivers of any pollen that happens to be blown or carried in, but as organisms capable of making decisions, choosing between potential mates on the basis of their suitability. And the mates are forced to compete with one another to gain acceptance, turning each stigma into a microscopic equivalent of the arena in which some male birds put on frenetic displays to impress the waiting females.[416]

Viewed in this light, trees take on new meaning. It becomes

more and more difficult to go on seeing and describing them simply as large vegetables. And more and more necessary to find new ways of redefining an individual organism and describing its real identity.

I have tried elsewhere to identify the biological origins of awareness. And been forced to conclude that contact, communication and recognition all take place at a very simple level – occurring even amongst social bacteria that seem to be able to recognise self from non-self.[398] But I have also become aware of a natural divide, of a distinction between those creatures that possess true individuality and those that do not. The difference is one of personality – and it brings with it a whole host of new talents and abilities, while also placing restrictions on some old ones. Many ants, termites and bees are so intensely social that it helps to think of them as diffuse organisms – as bodies with up to twenty million mouths spread out over several acres. Each such superorganism has the equivalent of a stomach, sense organs, reproductive parts and even the rudiments of a brain. It enjoys a kind of group identity which makes it possible for parts of the society to recognise each other and to take action against intruders who don't carry the proper chemical calling cards.

The beehive is not unlike a tree in that its anatomy is diffuse. Parts of it are quite capable of hiving off to form a new community, just as shoots and cuttings may be lopped off an oak and cultivated elsewhere. Each association is simple enough to do this, but complex enough to be able to respond as a unit, meeting environmental challenges in a meaningful way. Some, like the alders and wattles which send out warnings, may extend their sphere of influence well beyond traditional boundaries. But none have the kind of cohesion that creates a distinct sense of self-identity and an ability to recognise other associations as individuals in their own right.

Ants and plants sometimes seem so sensitive to others of their kind that they appear to be in an almost extrasensory contact. But our kind of complexity brings with it an inherent obstacle to such communication. We and the apes and the whales, most cats and possibly even some birds and fish, have acquired unique and independent personalities. And with them, certain liabilities. We have gained much, but seem in the process to have lost or concealed other lines of communication. A barrier has been erected, which is breached only in emergencies – when personality is suppressed by more interpersonal concerns and we revert to dealing with other individuals at fundamental levels. Which may explain why it has proved so difficult to establish reliable evidence for anything like telepathy in action between one human and another. Our personalities keep getting in the way.

People *are* different. Not only from each other, but from all those plants and animals which have not refined their persons into recognisable personalities. The change is qualitative rather than quantitative. It did not come about as a result of an increase in the number of apical meristems or a growth in brain or group size. Exactly how and when it took place remains a mystery, but the result is clear. The system now includes a number of species whose nature is relatively independent of their substance.

They consist of cells grouped into organs arranged into complex organisms, but somewhere along the line something else was added. A new and magic ingredient which led to the growth of individuality, the appearance of the person, and the existence of mind.

I am, therefore I think.

Chapter Three

PROCESS

Take tuberculosis.

Nothing personal, you understand. Just as an example of one of the most important and least acknowledged forces in nature.

During the last few decades, the mortality rates for tuberculosis have dropped dramatically. In England and Wales, for instance, 900 teenage girls died of TB in 1946. But by 1961, the recognition and isolation of infection, basic improvements in nutrition and hygiene, and the availability of specific drugs, had succeeded in reducing this number to just 9 dead girls. The picture on the planet as a whole is far less rosy; but there is something in the figures, in the pattern of spread and the frequency of infection during the last century, that ought to give us pause for thought.

Some diseases, such as smallpox and polio, are so devastating that they seem not to have played much part in medical history until populations became large enough to support them. Either they never appeared at all, or if they did, simply wiped out small isolated communities before vanishing again without trace. But right from the beginning, tuberculosis was different. It hung around.

There is skeletal evidence of the effects of TB in graves from Germany dating back 10,000 years – and similar evidence from the Old Kingdom in ancient Egypt. Engravings from 2500 BC show deformities of the spine which, together with descriptions in the hieroglyph, provide a clear diagnosis of death by consumption.[254] In ancient China, the disease was common and recognised as *laoping*. In India, they called it *yakshma*. Hippocrates described TB in Greece during the fifth century BC; and reading between the lines of the New Testament, it appears at least likely that Christ himself was consumptive.[296]

At first the effects of tuberculosis were scattered, but as people moved more often, travelled more widely and came to live in larger communities, these became epidemic. Waves of TB spread across the world, devastating cities, bringing a new "white plague" hot on the heels of the "black death". The last and greatest European epidemic began in England during the sixteenth century and reached a peak of mortality in London around 1750. Each of the capitals of western Europe was affected in its turn during the late eighteenth and early nineteenth centuries, peaking in the cities of eastern Europe by about 1870. And then something peculiar took place. There was a sudden, marked and inexplicable decline in TB everywhere that records have been kept – beginning, it seems, in Germany in 1882.[139]

It is possible that TB is one of those diseases that is self-limiting. It has a violent effect on communities, like isolated Amazonian tribes, to whom it is entirely new – and far less dramatic effect in areas where people have had the time to develop some resistance to it. Over 90 per cent of people in the European industrial countries early this century had been infected with TB as children and conquered the disease, but this acquired immunity cannot, on its own, account for the sudden decline in TB at the end of the last century – when rapid industrialisation and urbanisation should have been leading instead to a marked increase in the disease.

It has been suggested that we may be developing a more resistant population, but this too seems statistically unlikely. The removal of what amounts to less than 1 per cent of the population could not significantly affect the resistance of the remaining 99 per cent. A mortality rate of around 90 per cent, removing all those liable to catch TB, would be needed to bring about such a change. All recent controlled mini-epidemics – those taking place in relative isolation on ships or in boarding schools, where there is 100 per cent contact with the disease – have produced up to 20 per cent morbidity. This is an infection rate as high as any that prevailed during the eighteenth century and a clear demonstration that we are *not* now more naturally resistant.[423]

Surgical intervention in cases of tuberculosis did not begin until 1912, and chemotherapy (mainly with streptomycin) was unknown until 1944. So it was not any dramatic advance in medical treatment that led to the spontaneous decline in the late nineteenth century. But something did happen in Germany in 1882 that could be very significant.

In 1876, there was an outbreak of anthrax amongst cattle in Silesia. A country doctor working near Breslau took samples from the spleen of infected cows and succeeded in transferring the disease to several mice, all of whom produced specimens of the same bacteria. Dr. Robert Koch was convinced that these micro-organisms were the cause and agent of the dreaded cattle disease and his paper on its life cycle led both to a teaching post at the local university – and to growing fame as the founder of modern medical microbiology. Koch went on to track down the organisms responsible for cholera, bubonic plague and sleeping sickness. But his greatest triumph, the one for which he was later to receive the Nobel Prize, was the description of *Mycobacterium tuberculosis* – the tubercle bacillus. And he made this discovery in 1882 in Germany, announcing it triumphantly on March 24th before the Physiological Society of Berlin.

Almost immediately there was a marked decrease, not only in the incidence of the disease, but also in its mortality. Deaths fell from 600 per 100,000 to around 200 in less than a decade – and they have continued to slide to the present level of about 20 out of every 100,000 people infected. The recent improvements can all be attributed to better medical care, but nothing comparable happened to account for the sudden and rapid decline which is evident in Hamburg and Berlin during the 1880s. Nothing, that is, except Koch's discovery and a spreading awareness of what lay behind the disease that had come to be called "Captain of all the Men of Death".[119]

I am well aware that I stretch the established facts with this suggestion that just knowing the reasons for something might help to make it better. There is still clearly room for other more prosaic explanations, but I have a suspicion that this is at least an option we ought to take seriously. It is possible that things do indeed work that way – that knowledge, even before it has a chance to be applied, can sometimes have a power and an influence of its own.

There are other examples of this fascinating effect in action.

Resonance

We live on a magnet. No one knows exactly why, but it seems to have a lot to do with our planet's core of molten iron and nickel. The field, which appears to be created by these interactive metals, is similar to that which would be produced by a powerful bar magnet lying at the centre of the Earth and pointing approximately north and south. But there the resemblance ends, because it is anything but constant.

Geomagnetism varies in two major ways. One is a long, slow change, which has brought about a 5 per cent reduction in field strength over the last hundred years. This is known as secular variation and its cause remains mysterious, but it seems to follow a great cycle driven from somewhere in the centre of the galaxy. The other changes are more rapid, producing short-term fluctuations that can be linked in many instances with events in the Sun.

Our star has a magnetic field which also fluctuates, surging to a peak every eleven years as thermonuclear reactions bubble up from its incandescent core, spotting the surface with visible blemishes and buffeting us with streams of charged particles borne across space in the solar wind. In addition to this sunspot cycle, we accentuate solar rhythms ourselves by moving towards and away from our energy source once each year in an elliptical orbit – and by coming under the disturbing tidal influence of our own satellite once every 29 days 12 hours and 44 minutes of the lunar month.

This already complex syncopation has been even further compounded by discovery during the 1970s of a division of the Sun's magnetic field into sectors which stretch from pole to pole like the sections of an orange. The field in each sector is oriented in the opposite direction to the sectors on either side and, as the Sun turns, it presents a new region of its magnetic face to us once every 8 days, producing a flip-flop in polarity with a day or so of turbulence in between.

At our end of this cosmic flutter, Earth has perturbations of her own. Our local magnetic field interacts with the sun-charged particles in the atmosphere to produce a sensitive electromagnetic field. The earth surface and the ionosphere act as the charged plates of a condenser, producing an electrostatic field which pulses in the low frequency range, and an electrodynamic resonating cavity that hums along at roughly ten cycles per second. In addition, every flash of lightning in this space releases a burst of radio energy which bounces between the poles, travelling back and forth along the lines of Earth's field several times before it peters out. And, we now learn, there are also large direct currents within the fabric of Earth itself, constantly circling the restless planet, creating their own subsidiary electromagnetic fields.

Our lump of rock is anything but inert. It sizzles and crackles with energy; pulsing, itching and breathing like a living thing; responding directly to changes in itself and in the environment. And we ride these waves like veteran surfers, dealing instinctively with their fluctuations, anticipating changes in frequency which lie beyond the limited scope of our usual sense organs. We learn to read between the lines. We resonate in natural sympathy with our planet.

Think, as I suggested fifteen years ago, of an insect sitting on a tuning fork.[396] It has no ears, but is nevertheless aware of a vibration in its perch produced by striking a second and similar fork some distance away. The two metal instruments resonate in sympathy with each other and the insect becomes sensitive to the occurrence of events beyond its normal ken. It is in this way that subtle stimuli, too small to make any impression on the usual senses, are magnified and brought to our attention. This is how things here are sometimes able to respond to each other in extraordinary ways – how the apparently supernatural can be eased into the more comfortable confines of natural history.

Few people in the last decade have proved more adept at such reconciliation than the biologist Rupert Sheldrake.

In 1981, Sheldrake published what he called *A New Science of Life*.[341] The book was promptly condemned by one of the most influential scientific journals as a prime candidate for burning – sufficient in itself to alert those of us who have learned from Thomas Henry Huxley that "it is the customary fate of new truths to begin as heresies."[180] What outraged the conservative editors of *Nature* was that Sheldrake was insufficiently respectful of the sacred disciplines of genetics and molecular biology and had the impudence to insist that an additional causal principle was necessary to explain the development and organisation of matter into the extraordinary variety of plant and animal forms. He called his principle "formative causation" and further offended the high priests by suggesting that it operated through "morphogenetic fields" which did not themselves consist of energy or matter and were unaffected by space or time.

It is easy to understand the outrage, but it is not that easy to dismiss Sheldrake's ideas. His thesis is both rigorously scientific and well capable of being tested. And in the tests which have been made, it seems so far to be sound. He says, in essence, that it is easier to do something which has already been done – even if there is no direct physical link between the two attempts to do it. And he predicts that the principle will apply equally well to the selection of a particular physical form by a substance in the process of crystallisation, or to the success of rats and children in learning a particular task.

Common crystals like sodium chloride have been forming for thousands of millions of years and readily take their characteristic cubic form, but it is not always easy to force a new liquid substance into its appropriate solid state. Ethylene diamine tartrate, for instance, was first produced more than thirty years ago as an organic version of a substance with the useful talent of seeking out and locking up unwanted minerals, like the metallic ions in hard water. The company making it eventually found a way to produce the solid form and for a while successfully marketed large and perfect crystals. Then suddenly their tanks started growing a new form – the monohydrate version of the chemical – which was not only useless to them but soon spread to other plants. "During three years of research and development, and another year of manufacture, no seed of monohydrate had formed. After that, they seemed to be everywhere."[167]

The orthodox explanation for the fact that novel substances crystallise ever more easily, is that seeds infect subsequent solutions and start the crystals growing there too. But there are examples of such infection taking place over vast distances and getting into even sealed samples of the solution.[398] Sheldrake insists that the crystallisation of a substance is facilitated by the mere fact that it has crystallised before – and the more often it is crystallised, the stronger this influence should become, because increasing numbers of past crystals contribute to the morphogenetic field responsible for creating such a form. In other words, crystals take a characteristic form because they resonate in sympathy with others of their kind. Like begets like, even at a distance in space and time.

One of Sheldrake's tests of morphic resonance in action amongst humans is even more elegant.[399] He persuaded a Japanese poet to provide him with three short rhymes in Japanese, all with a similar sound structure. One of these was meaningless, just a jumble of unconnected words. Another was a freshly composed verse, and the third was a traditional rhyme known to generations of Japanese children. Nobody involved in the experiment understood Japanese or knew which was which, but all found that English-speaking subjects were able to memorise one of the three rhymes far more easily than the other two. This one was, of course, the popular nursery rhyme – made more accessible, according to Sheldrake, by the fact that its morphogenetic field was built up to a very powerful level by constant repetition over hundreds of years.

In trying to understand Sheldrake's thesis, I find one of his own analogies most helpful. Think, he suggests, of the first meeting between a Martian and a television set. The initial and logical reaction of any intelligent alien unfamiliar with our technology might be to assume that the box actually contains the little creatures whose images appear on its screen. When this theory is disproved by opening the box – it contains nothing but transistors, condensers and wire – it might be equally proper and no less logical to adopt a second and more sophisticated hypothesis; one which holds that the images must be produced by complicated interactions amongst the visible components. It could take a while, and would certainly require a considerable intuitive leap, to carry the Martian through to the ultimate conclusion that the images in fact depend on invisible influence from a distant transmitter.

We are at the moment in the same position as that Martian, poised on the brink of the third theory, finding it difficult to make the necessary leap, unwilling (and in some cases totally unable) to go beyond what science can actually see. The hypothesis of formative causation makes the jump, appreciating the importance of genetics and biochemistry, but recognising also that there may be outside influences – other organisms broadcasting information on an appropriate wavelength, sharing knowledge, putting out the news.

The possibility of such resonance between things is an important idea. It could explain a great deal that still remains mysterious, perhaps even providing a mechanism for the direct action of knowledge about something like tuberculosis on the incidence of the disease.

Richard Dawkins, the Oxford biologist who drew attention to the tendency of genes to look after their own interests, was also responsible for identifying the "meme".[79] He points out that words, phrases, fashions, theories and ideas are in effect non-physical genes, a kind of abstract DNA. They are very much alive, passing from brain to brain, propagating themselves by imitation. Looked at in this way, ideas have a reality of their own. They are realised physically, over and over again, as actual structures and processes in human nervous systems. And if Sheldrake is right, each time they form they gain strength and momentum, building up their morphogenetic fields, becoming an evolutionary force in their own right.

In the struggle for survival, the most successful memes will be those that either fit the facts or meet a basic need. And those that can serve both functions are likely to have a special vigour. When Robert Koch described the tubercle bacillus, he not only added to our sum of knowledge, but actually altered the TB meme in an important way. It mutated under his influence from an amorphous and imaginary terror to a slender and tangible bacterium – one of a group of tiny things that can even be destroyed by sunlight. This alteration gave the meme new and popular momentum, putting it in line for a Nobel Prize, but it seems to have done the germ no good at all. Old ideas die as easily as new people.

Darwin never considered the evolution of ideas. But these run the gauntlet of natural selection in much the same way as any physical unit. There is the same struggle for existence, the same success for those best fitted to survive. And many of the same mysteries about the mechanism involved.

Natural Selection

It is impossible to describe the difference between male and female day-old chicks – and yet there are those who can tell them apart.

The female and male genitalia in birds of this age are indistinguishable, even to trained biologists. There are no apparent distinctions in size, weight, body shape, voice pitch, beak length or skin and feather colour – nothing about the chicks that gives any obvious clue to their sex. Nevertheless, there are people who can separate the sexes and who earn substantial salaries from an industry that has no interest in wasting money on rearing unproductive male poultry.

Most of these experts are Japanese, so I went recently to visit the centre near Osaka where many of the best chicken-sexers are trained. I was hoping, as a biologist, to learn something from them, to uncover the secret cues. But I found no obvious answer. All arts in Japan are learned by example, by living with and working alongside a *sensei*, a teacher, for long enough to master the art – and chicken-sexing turned out to be no different. Novices learn by looking over the shoulders of experienced workers, and go on doing so until they acquire the skills, almost by a process of osmosis. The experts themselves cannot explain how they do it. But they do, and so eventually do the trainees, without hesitation and with a success rate of over 99 per cent.

It is clear to me, from what I saw in Japan, that no conscious effort in sexing could ever approach such phenomenal speed and accuracy. Skills of this order seem to be acquired implicitly and unconsciously – and are actually hindered and blocked by any attempt to exercise our ability to reason. They rely instead on the presence and application of what could be called an intuitive edge. Successful chicken-sexing is a little like trying to recall a name or a dream that, for the moment, escapes you. The harder you try, the less successful you are likely to be.

At Brooklyn College in New York, psychologist Arthur Reber is studying the nature of implicit and explicit knowledge. In one test, he provided two groups of students with identical lists of nonsense words that had been composed very carefully with new and arbitrary rules of grammar. Those in the first group were asked to analyse the words in an attempt to discover the grammatical rules. Those in the second group were asked only to memorise the list. Both groups were then given some additional words and asked whether these did or did not obey the same rules. Reber discovered that the second group, the nonanalysers, gave far more correct answers. He concluded that "subjects who approached the task from a neural stance learned more about the synthetic grammar than did those who consciously tried to decipher the rules."[297]

It seems that when faced with particularly subtle tasks, people who feel or intuit their way through them actually have a competitive edge over those who consciously try to think their way through. It may be true that in time, with the aid of appropriate instruments, we will be able to isolate and identify the cues responsible for successful chick-sexing. But in the meantime, there are people doing it very efficiently without such understanding, making decisions without knowing why, and there is considerable historical evidence to show that the exercise of this ability can be highly creative.

Karl Friedrich Gauss is sometimes described, along with Archimedes and Newton, as one of the three greatest mathematicians of all time. He was a prodigy, obsessed with numbers, making several remarkable discoveries (including the method of least squares) while still in his teens. He worked out a non-Euclidian geometry, a method for constructing an equilateral polygon of seventeen sides, and proved the fundamental theorem of algebra before his twenty-second birthday. But the proof he really wanted, the fundamental theorem of arithmetic, continued to elude him. It wasn't until he was all of twenty-four, that he discovered a way of proving that every number can be represented as the product of primes in one and only one way. And the answer came out of the blue. After years of struggle, he recalled in his diary: "Finally two days ago, I succeeded, but not on account of my painful efforts. Like a sudden flash of lightning, the riddle happened to be solved."

Much of our success as a species is due to the deliberate and conscious application of explicit knowledge, but there is no denying the power and creativity of the unconscious. We seem to have an ability to know what to do in complicated situations without being able to explain how or why. We act on impulse, on a hunch, making snap decisions that very often turn out to be the most appropriate. And we may not be alone in this. There are signs that other species have similar talents and that intuition and a sense of what is appropriate have played a major role in evolution.

To appreciate the complexities of evolution in action, Fred Hoyle suggests that we consider the problems of a blind man confronted with one of those infuriating multicoloured Rubik's Cubes.[171] The man cannot see the result of his manipulations. He has no way of knowing whether he is getting nearer the solution or whether he is scrambling the cube still further. He acts at random and his chances of producing a perfect and simultaneous colour matching for all of the cube's six faces is about 50,000,000,000,000,000,000 to 1.

But suppose that one of those teenage prodigies who can un-scramble a cube in less than 23 seconds is available – and agrees to stand behind our beleaguered blind man. At each move of the cube, the young expert remains silent as long as the move is in the right direction. But if a move does not advance the cube towards its solution, he simply and quietly whispers "No", and the blind operator reverses the move just made and tries another one – and goes on doing so until the observer says "Stop". If one minute is allowed for each successful move and an average of something like 120 moves are needed to reach the solution, even a blind man can do the task in a comparatively short time. The presence of the observer and the use of that one short word "No" at appropriate times, makes all the difference between a directed solution that takes just two hours – and a random one that could take 300 times the age of Earth.

The orthodox view of evolution is a little like this situation with the Rubik's Cube. Mutations, which occur at random and in unpredictable directions, are represented as moves made by a blind man. And natural selection, as it is exercised by the environ-ment, is seen to operate on a mutating species in much the same way as an observer who decides whether or not the moves it makes are good. But the analogy is incomplete and misleading, because even neo-Darwinian evolutionists insist that natural selection is unintelligent – it does not know the solution in advance. All it can do is make limited value judgements about isolated moves.

"Natural selection," in Darwin's own words, "is daily and hourly scrutinising throughout the world, the slightest variations, rejecting those that are bad, preserving and adding up all that are good, silently and insensibly working . . ."[73] The key word here is "insensible". Natural selection is *not* an expert observer. It is a considerable force, but an unintelligent one, in its own way as blind as the operator of the cube or the pattern of mutation. And yet evolution goes on getting things right, guiding mutation in produc-tive directions, making choices that result in the right solutions, producing appropriate changes in an incredibly short time.

There is something about it that flies directly in the face of the careful discrimination which led Patrick Matthew in 1831 to coin the term "natural selection" as something distinct from "artificial selection", which he described as selection directed by an outside intelligence. He was thinking of course of a human agency, but the more closely one looks at what is now described as natural selection, the more it seems to include an essentially artificial component – something that keeps pushing it in a particular direction. It is abundantly clear from the fossil record that when organisms do change, the modifications which occur are of a kind which improve fitness far more often than can be expected from changes taking place on a purely random basis. And, what is even more significant, is the fact that such changes quite often occur before the need for them arises.

A good example is the development of the amniotic egg which made it possible for backboned animals to migrate completely from water to land. This was no minor adjustment. The eggs of fish and frogs are simple blobs of jelly that must be kept in water to survive, but reptile and bird eggs have an amnion – a sac filled with fluid that floats each embryo safely in its own private pool and allows the parent responsible to wander Earth at will. And this stunning advance, as big in its way as the appearance of feathers or the development of limbs, was apparently *not* made in response to the pressures produced by a life on dry land. Amniotic eggs in protective shells were, it seems, being laid even by mesosaurs – totally aquatic dinosaurs that never left the water, but carried within them both the genes and the means for doing so. Reptiles and their eggs appear to have been wonderfully pre-adapted for life on land.[61]

There are similar scenarios involved in the transition from walking to climbing. The most alluring of all amphibians are the tree frogs – almost five hundred species of pop-eyed, multi-hued clowns and acrobats that hang by their toes from twigs in the rainforest or balance lightly on swaying reeds beside most of the world's great rivers. They range from half-inch grass frogs *Hyla ocularis* in the cypress swamps of Florida to hand-sized *Hyla maxima* which press against tree trunks in the Amazon – looking, despite their large size, still too small for their loose enveloping folds of skin. But all tree frogs have the same kind of feet with terminal suction cups, elongated fingers lined with unusual connective fibre, and an extra segment of cartilage pad shaped like a built-in climbing boot. All of which makes sense for arboreal animals, but it has now been discovered that similar adaptations exist amongst several families of terrestrial frogs which never climb and never have climbed trees.[269] Apart from occasional leaps, for which they are very well designed, these species tend to be confined to the ground, but they have the look and the equipment of frogs that are getting ready to be climbers. They have been armed by evolution for a pre-emptive strike at life in the trees, should such a strategy ever become necessary. They are pre-adapted to a new habitat before its pressures have been brought to bear – and there is nothing in the traditional concept of natural selection which allows them to have such an advantage.

Nothing, that is, unless it is admitted that there are connections between all parts of the environment. Creative tensions that allow species in one part to be fed with information that is relevant to the whole. It becomes necessary to assume that the environment within which such predictive natural selection acts, is in some sense intelligent and exercises a degree of pattern and control. Unintelligent selection can only produce random and unintelligent results. But if the system itself has an inherent form, and encourages developments which are appropriate to that form, then fitness for a species or a new mutation begins to mean something more than a slight statistical shift in the odds on its survival.

All that is necessary is that the environment, or some observant part of it, should whisper "No" at the critical moments of poor choice, forestalling whole sequences of unproductive moves. This guidance can be as trivial as the unconscious ability that makes it possible for someone to sex day-old chicks, or as profound as the sequence of specific mutations that led, against all the odds and in an incredibly short space of time, to the production of whole haemoglobin or the perfection of the vertebrate eye.

I suggest that natural selection is subject to such guidance, that it is not altogether neutral or unintelligent, but can respond to a still small voice that, in effect, says, "No, that doesn't feel right. Try again in another way." And goes on nagging just often enough and long enough to make a powerful difference. Biologists talk a lot about survival of the "fittest", using the word more often than not in its adverbial sense as a description of an organism which is big enough, strong enough, smart enough – in some way brought into a condition suitable for meeting the demands of the environment head on. This interpretation probably goes all the way back to the Old English origin of the word *fitt* as a description of an array of fighting men drawn up for conflict. But it seems to me to be more revealing to consider the transitive verbal form of "fitt", which concentrates instead on those properties of an organism which give it a proper measure, adapting it to, and putting it in harmony with, the environment. "Fitness", in this sense, means befitted to and not armed against.

It may sound passive and unheroic, lacking in the muscular qualities we have come to associate with the struggle for survival, but true fitness is a very powerful property indeed.

Fitness

The success of science during the last three centuries has left us with an exaggerated respect for its method, which emphasises information, logic and reason. Much of our knowledge has, it is true, been accumulated by the patient collection of facts and a cautious modification of existing theories. But, as Philip Goldberg points out in a rewarding recent book on the powers of intuition, all the great revolutions in thought have come about as a result of creative leaps and soaring flights of imagination that owe little to direct intellect.[127]

When Albert Einstein published his General Theory of Relativity in 1915, it superseded Newton's classic concept of gravity with a new view that allowed grand conclusions to be drawn about the universe as a whole. The idea grew naturally out of his Special Theory of Relativity, drawing on this earlier insight and decorating it with the fruits of a decade of further understanding, but the actual breakthrough came only with what Einstein later called "the happiest thought of my life". It was a stray, and at the time illogical, vision of a person falling from a roof and the realisation that someone in that position was both at rest and in motion at the same time. But this flash provided the creative impetus necessary to change the way in which we all still look at the world. "There are no logical paths to such natural laws," Einstein said of his discovery, "only intuition can reach them."[94]

Isaac Newton's own insights seem to have depended on similar flights of pure imagination. The economist John Maynard Keynes says of him:

It was his intuition which was pre-eminently extraordinary. So happy in his conjectures that he seemed to know more than he could have possibly any hope of proving. The proofs were dressed up afterwards; they were not the instrument of discovery.

And this seems to be true of all branches of science. "Mathematics," in the words of a modern adherent, "is a series of great intuitions carefully sifted, refined, and organised by the logic men are willing and able to apply at any time."[207] But the truth is that they usually do so later, in easy retrospect.

It is the fruits of such hindsight, all carefully ordered and logically arranged, that we see in the final published papers. Theories there seem to flow with smooth inevitability from lucid sequences of well-planned research. But science is seldom like that at all. It is more usually a matter of side-trips, false starts and dead ends – a haphazard journey interrupted by mishaps and frequently abandoned altogether as a result of poor equipment or sloppy technique. More often than not, it is a maze from which we only emerge with any credit as a result of vague hunches and gut feelings that put us back on the right track at critical moments. And the more profound the final insight, the greater it seems, the vital role of intuition. As the philosopher of science Karl Popper says, "There is no such thing as a logical method of having new ideas, or a logical reconstruction of this process. Every great discovery contains an irrational element or a creative intuition."[281]

The nineteenth-century French mathematician Henri Poincaré got off to an inauspicious start. His eyesight was poor, there was something wrong with his motor coordination, and schooling interrupted by the Franco-Prussian War left him regarded as something of a retard. But he had a special gift, an innate ability to perceive patterns whose elements, in his own words, "are harmoniously disposed so that the mind without effort can embrace their totality . . . divining hidden harmonies and relations."

This sense of what was elegant and right, served Poincaré well. On one occasion, he recalled, "The changes of travel made me forget my mathematical work. Having reached Coutances, we entered an omnibus to go some place or other. At the moment when I put my foot on the step, the idea came to me, without anything in my former thoughts seeming to have paved the way for it, that the transformations I had used to define the Fuchsian functions were identical with those of non-Euclidian geometry. I did not verify the idea; I should not have had time, as, upon taking my seat in the omnibus, I went on with a conversation already commenced, but I felt a perfect certainty." There was no further need for proof or substantiation. Poincaré *knew* he was right, but his memoir of the occasion ends with the throwaway line: "On my return to Caen, for conscience's sake, I verified the result at my leisure."[325]

This certainty is characteristic of true intuition. The answers come with what psychologist Jerome Bruner calls "the shock of recognition".[35] They come suddenly and surprisingly, but fit so well that when the surprise wears off, we are left thinking, "Of course. It is obvious. How could I not have seen it all along?" And from that point on, the missing piece slots neatly into place, the picture is complete, the puzzle is solved and it is hard to remember what it felt like *not* to know the answer.

And this process of discovery is by no means unique to science. Mozart, in a letter to a friend, described his creative gift as one coming from outside himself.

> When I am, as it were, completely myself, entirely alone, and of good cheer – say, travelling in a carriage, or walking after a good meal, or during the night when I cannot sleep; it is on such occasions that ideas flow best and most abundantly. Whence and how they come, I know not; nor can I force them . . . Nor do I hear in my imagination the parts successively but I hear them, as it were, all at once . . . The committing to paper is done quickly enough, for everything is already finished; and it rarely differs on paper from what it was in my imagination.[127]

This enviable flow of inspiration, fully formed, was Mozart's great glory – the result, it seems, of an unusual ability to sustain the intuitive moment beyond the brief flash that leaves most of us blinking and fumbling for answers that were clear in the moment of illumination, but seldom last long enough for us to put them into words or get them down on paper.

Bach had some of Mozart's flair. "I play," he said, "the notes in order, as they are written. It is God who makes the music." Milton

73

wrote that the Muse "dictated" to him the whole "unpremeditated song" that we now know as *Paradise Lost*. Robert Louis Stevenson dreamed the plot of *Doctor Jekyll and Mister Hyde*. Samuel Taylor Coleridge awoke with what he called "a distinct recollection" of the whole of "Kubla Khan", which he wrote down without conscious effort, pausing only when interrupted by the infamous Visitor from Porlock. By the time that Coleridge returned to his room, the end of the poem was lost for ever. It had "passed away like the images on the surface of a stream into which a stone has been cast." The flow was broken and the work remains tantalisingly incomplete.

The onset of such illumination has characteristic symptoms. We become subject to "cold chills", "tingles", "burning sensations" and "electric glows". We get "gut reactions" and "feel things in our bones". The reactions are visceral, but often have superficial symptoms. The poet A. E. Housman remained resolutely cleanshaven. "Experience has taught me," he said, "when I am shaving of a morning, to keep watch over my thoughts, because if a line of poetry strays into my memory, my skin bristles so that the razor ceases to act." Creative ideas are often preceded by intimations, by fuzzy feelings that something is about to happen.

These are well described by philosopher Graham Wallas as "a vague, almost physical, recurrent feeling as if my clothes did not quite fit me".[393]

There is, however, no mistaking the moment of sudden breakthrough. Psychologist Rollo May sees it as "a special translucence . . . The world, both inwardly and outwardly, takes on an intensity that may be momentarily overwhelming . . . I experience a strange lightness in my step as though a great load were taken off my shoulders, a sense of joy on a deeper level that continues without any relation whatever to the mundane tasks that I may be performing at the time."[240]

There is something too about the truly creative inspiration that distinguishes it from stray and misleading hunches. It has a certain symmetry, a kind of inevitability that gives it beauty. It fits. English physicist Paul Dirac in 1930 predicted the existence of antimatter – of particles of equal mass and opposite charge to the electron and the proton. He did this without physical evidence or proof of any kind, but with inspiration and a set of equations of such natural elegance that they earned him a share of the Nobel Prize in 1933.

American physicist Richard Feynman won his portion of the same prize thirty years later with discoveries about the behaviour of electrons that he was able to describe in mathematics of such heart-wrenching simplicity and grace that their truth is evident even to those like myself who are far less numerate.

Simplicity of this kind seems very often to be a key to the aesthetics of truth. The best answer is usually the simplest one consistent with the facts. The one that fits without qualification or justification. It just slides into place without effort in a way that is not only satisfying, but truly beautiful. This, I think, is how directed evolution has to work. "We leap," in the words of psychologist Donald Norman, "to correct answers before there are sufficient data, we intuit, we grasp, we jump to conclusions despite the lack of convincing evidence. That we are right more often than wrong, is a miracle."[176] Or perhaps a natural consequence of the fact that we are tuned in to the world on wavelengths that are sensitive to harmony and truth. We have a predisposition to prefer that which fits.

In a certain limited sense, perceived fitness has something to do with familiarity and expectation. I have, on several occasions, passed people I know well on the back streets of Bali or Kathmandu and failed to respond to them for several seconds simply because I wasn't expecting to see them there, so far from their normal environments in London or Los Angeles. The usual visual cues for recognition were as good as ever, and at some level I registered these, but social manifestation of this recognition was delayed and inhibited by the circumstances – until I was forced to double back and say a belated hello. There is a similar situation, nicely quantified by an experiment at Acadia University in Nova Scotia. Students there were shown a series of photographs and asked to select from these the particular object which was most familiar to them. The vast majority chose pictures of things such as their Student Union Building, despite the fact that each test series included a full colour, clandestine close-up of the back of each subject's own hand.[421]

The saying "I know it like the back of my own hand," is true in the sense that this object is probably the most frequently occurring retinal image in our lives. But it turns out not to be true in that the image becomes unfamiliar and remains unrecognised in an unexpected set. The shock recognition of an equally unexpected intuition is obviously something of a different order. It strikes home, having impact and moment no matter how unfamiliar it might be. Even if it is something you have never heard or seen or dreamed of before, you recognise it instantly, knowing that it is right and appropriate, appreciating that it fits.

This response is so basic that I have no hesitation in suggesting that it is something universal, independent of intellect or experience, and available perhaps to other species in much the same way that it is made manifest to us. And if this is so, we are dealing with a faculty with clear survival value, a biological attribute that gives decision-making a predictive quality. We are looking at the source of, or at least evidence for the existence of, that still, small voice that whispers "No" when we make a bad move, or simply says "Stop, that's it" when we hit upon the right solution.

That kind of knowing could make a lot of difference.

Perfect Speed

Jonathan Livingston Seagull rose above his personal problems with the discovery of "perfect speed".[4]

This is not exactly evidence drawn from natural history, and the concept of perfect speed is a simplification of understandings better described in oriental philosophy, but it is nevertheless one with some relevance for biology. There are circumstances in which we seem to be capable of performing beyond our abilities, in doing things and knowing things that are otherwise impossible or inaccessible to us.

Racing drivers, particularly those involved on the Formula One circuit, are super-fit athletes with a fine awareness of their own tolerance and the capacity of their machinery. They make a living out of pushing themselves and their cars to the limit, but most of them also recognise a level of performance which exists beyond normal bounds. They talk about such transcendence only with reluctance, but it occurs often enough to be part of most professionals' experience and to be known by them as "driving above yourself".

The phenomenon is rare and becomes manifest mainly at times of risk and sudden accident when there is no time to respond with logic or reason. When danger appears with horrid suddenness, such drivers switch, as it were, into overdrive and produce un- usually brilliant performances which avoid the source of the problem and bring spectators to their feet in astonished applause. The drivers describe the experience as one of total calm – "you know exactly where you are going, what line to take, and it feels as if you've all the time in the world."[111]

There is no question of the survival value of such an ability, though it is difficult to see how it could have been acquired in the normal course of evolution by natural selection. Travelling at over 200 miles an hour is not a part of most vertebrate experience. But it seems that we have a reservoir of spare capacity built in to our systems, an area of overdrive available to meet circumstances that are impossible to predict. A sort of evolutionary afterthought, something extra in case of emergencies.

There are numerous anecdotal accounts of people in times of crisis doing extraordinary things – of mothers lifting cars and trucks to free trapped children or of survivors leaping, running or swimming across life-threatening hazards in impossible ways. These accounts are by their very nature, difficult to control or confirm, but there are documented athletic records of similarly superhuman feats – none more extraordinary than the "mutation performance" that took place in Mexico City on October 18th, 1968, the day that, in one sports writer's words, "it all came together for ever".

It was a cool day with a slight following wind. Air resistance at an altitude of 7400 feet was reduced below its sea level value. The track was equipped with a new "tartan" surface. And the athlete was a tall black man with the long legs of his race and muscles containing enzymes capable of short bursts of exceptional mech- anical force. All this was known, but still fails to account for the fact that Robert Beamon on that day eclipsed the world long jump record by an incredible 21.5 inches.

Ernst Jokl, professor of neurology at the University of Ken- tucky, calls Beamon's quantum leap "the greatest single feat in the recorded history of athletics" and estimates that it stands 84 years ahead of its logical time in history. In the years before Beamon's flight, less than 9 inches had been added to Jesse Owens' 1935 record distance of 26 feet 8.25 inches. "In one sense," Jokl admits, "it is entirely inexplicable. Nothing justifies the assumption that we are going to see another 29-foot jump, ever." Other experts agree and suspect that such a jump might be at the limit of human capability because the forces necessary are "on the verge of those which tear muscles and break bones". Beamon's own perception of what happened, is hazy. "I was frightened, man. I figured the pressure was on me. I was between time and space." And so he went out, hit perfect speed, and jumped 29 feet 2.5 inches.[173]

It is known that altitude has a direct effect on human physiology, lowering reflex thresholds and giving everyone sharper responses. It also seems to increase the sensitivity of that part of the brain that controls consciousness and muscle tone. When told the distance of his leap, Beamon suffered from a cataleptic seizure. He collapsed. His legs gave out altogether and, though he remained conscious, he could not rise again for a minute or more. Everything points to the conclusion that he was able, for a number of reasons, to draw on extraordinary reserves. He was the right person in the right place, but it also had to be the right time. Since 1968, another American athlete has cleared 28 feet on several occasions, coming once within 5 inches of Beamon's mark. But Carl Lewis has so far failed to bring together all the circumstances necessary to achieve the state of grace that seems to be required – not just to break a record, but to top a performance that was almost apocalyptic.

It is interesting that Beamon felt himself to be somewhat detached, "between time and space", at the moment of his jump. And that disciplines which actively encourage this condition as a method of transcending the limits of the self, originate at altitudes much like those of Mexico City. The *Yoga Sutras* of Patanjali give explicit instruction in the routes of meditation that lead through transcendence to *siddhis* or supernormal powers. A prerequisite for such powers is a state of awareness, a kind of total clarity with the whole nervous system working at maximum coherence, which translates as "being full of truth".

My reason for this diversion into Eastern mysticism is to provide a connection between the ability of organisms to live beyond themselves under certain circumstances, and the intuition which enables some of us, in Wordsworth's phrase, "to see into the heart of things". Mathematicians, in particular derive information about aspects of reality which have never been observed and predict consequences which were previously unsuspected. They operate as though the world itself were held together by the same intelligence that flows through each of us, abolishing the distinction between "outside" and "inside". The Taoist sage Lao Tzu said, "I know the way of all things by what is within me." And Einstein is described by one of his biographers as arriving at results by a "phenomenal intuition of what they should be by a deep inner contact with nature".[21]

All of which leads me inexorably – and not without a certain concern, because I remain a Western-trained scientist with all the pragmatism that entails – to the conclusion that living things are part of some larger process. And that they can, under certain circumstances, exploit that relationship in ways that give them access to supernormal powers – making them better fitted for survival. And proof of this ability is, I suggest, evident in some peculiar aspects of animal behaviour.

In August 1914, Private James Brown of the 1st North Stafford-shire Regiment was sent to join the Great War in France. On 27th September, his wife wrote to him with the bad news that Prince, his favourite Irish terrier, was missing. He wrote back: "I am sorry that you have not found Prince. You are not likely to. He is over here with me." The dog, it seems, travelled over 200 miles through the south of England, succeeded somehow in crossing the Channel and then negotiated another 60 miles of war-torn French country-side to the one set of trenches on the front line near Armentières, where its master could be found.[143]

Things like this keep happening. When Robert Devereaux was taken to the Tower of London in 1600, accused of plotting against Queen Elizabeth, one of the conspirators arrested with him was Henry Wriothesley, the third Earl of Southampton. Within weeks of his incarceration, the Earl was joined in his cell by a black and white cat with an ornate collar, that had apparently travelled a hundred miles up from the south coast and climbed down a chimney in the Tower to be reunited with its master. There is a fine portrait of the two of them, still together, in the collection of the Duke of Buccleuch, Captain of the second Queen Elizabeth's Body Guard.[244]

It is important, of course, to be sure that the animals involved were indeed the same ones and not coincidental look-alikes. Psychologist Michael Fox of Washington University records two cases of travelling cats that could be recognised by distinctive marks. One belonged to a New York veterinarian who left his pet behind when going to take up a new post 2500 miles away in

California. Some months later, an identical cat walked into his new home. Incredulously, he examined the animal and found a deformation on the fourth vertebra of the tail – an injury he had himself treated when the cat was bitten as a kitten. The second restless cat was a Persian called Smoky, distinguished by a tuft of reddish hair beneath its chin. Smoky leaped from a window of the family car when its owners were in the process of moving house. Neighbours near the old home in Oklahoma saw the cat prowling around there for several days, but a year later it turned up 300 miles away at the family's new and totally unfamiliar home in Tennessee.[118]

No biologist is surprised by stories of animals making their way, often across enormous distances, to established or traditional home sites. There is a vast literature on homing ability and an astonishing array of sensory cues known to assist organisms in the process of navigation. But what is different and disturbing about the anecdotes above is that they deal with animals becoming reunited with people in places which are not only totally unfamiliar, but have no biological, geographical or historical features that could be useful to natural history. There is no known gene or established sense that can be invoked to explain such behaviour.

J. B. Rhine at Duke University in North Carolina called it "psi-trailing" and gathered together a collection of fifty-four spontaneous cases in which "an animal, separated from a mate or owner, followed it into wholly unfamiliar territory under conditions that preclude the use of a sensory trail." He satisfied himself that these cases were based on reliable information, had independent corroboration and involved animals that could be recognised by some clear character such as a name-tag or an unusual scar. And concluded that, while the vast majority of abandoned pets fail to find their erstwhile owners, a sufficient number do so in extraordinary circumstances for the phenomenon to be taken seriously.[301]

Rhine also drew attention to the possibility of what he called a "team effect". In a series of tests on a California beach, Rhine buried a number of small wooden target boxes at random under four inches of sand that was flooded with twelve inches of water as the tide came in. Raking of the sand and subsequent disturbance by water and wind made it impossible to detect the sites visually and unlikely that the targets left olfactory or any other cues to their precise whereabouts. But two German shepherd dogs were able, in a series of 203 trials, to locate the hidden boxes underwater with a phenomenal success rate of 38.9 per cent. They sat down, as they had been trained to do, right over the sodden targets – when the odds against them doing so purely by chance were of the order of a thousand million to one.[300]

Or at least they were able to do so, *provided that* they were accompanied by their trainer and were being observed, from a distance and out of sight and hearing, by someone who had buried the boxes and knew, by reference to a grid system of white tape markers, when the dogs were getting near their targets. Animals or trainers operating on their own are never so successful. Rhine decided that, in this experiment at least, "a kind of fusion between the two", some connection between humans and dogs, was responsible. He called it "a functional integration of man and dog mentalities".

Parapsychologists today, more than thirty years after those tests were made, are inclined to dismiss them as poorly controlled, complaining of "contamination" and "experimenter effects". It is true that the test design makes it impossible to decide whether the observer or the trainer or the dog was responsible for locating the targets, but in a very real sense, it does not matter. The important fact is that living organisms, in this case of two different species, are apparently able to exercise some kind of faculty which still defies analysis – and to do so very successfully indeed.

I am impressed by the evidence, both anecdotal and experimental, for the existence of such unusual talents, which can justifiably be described as supernormal. Their evolutionary significance and survival value remain, in many cases, obscure. It is hard to decide whether they represent an archaic ability now in a process of decline, or some nascent quality peculiar to higher organisms and only now just beginning to show its true potential in human beings.

I am convinced that we can, in the right association or when it becomes necessary in a crisis, operate beyond our usual physical and sensory limits. The ability to do so seems to depend either on emotional factors that can carry us across individual thresholds, or on the sort of transcendence that takes place when "everything comes together". When it does occur, this switch gives us, and perhaps other species, something more than customary strength and awareness. And it can be very creative.

I intend in Part Two to try to look more closely at the origin, significance and potential of such talents.

Part Two

MIND

"One thought fills immensity."

WILLIAM BLAKE in *The Marriage of Heaven and Hell*, 1790

Earth, and everything on it, is under constant bombardment. We are battered by a ceaseless barrage of more than a hundred million impulses every second – a confusing avalanche of raw knowledge with which we cannot hope to deal.

The flood ranges from highly energetic cosmic rays, whose origin remains mysterious, but which at all latitudes have the intriguing property of coming rushing in largely from the west; through short wavelength gamma and X-radiation, which passes with relative ease through our bodies; to ultraviolet and infrared light waves that leave us with vitamin D and radiant heat; and the wide band of radio frequencies that bring us sound broadcasting, television, radar and a scattering of information about distant galaxies in collision.

Most of this news is irrelevant. It contributes nothing to an organism battling for survival on a much more limited field. To prevent being overwhelmed by the flood, we have evolved barriers which filter out the stuff we don't need. We funnel the confusion through what Aldous Huxley called the "reducing valves" of our brain and nervous system, so that "what comes out at the other end is a measly trickle of the kind of consciousness which will help us stay alive on the surface of this particular planet."[179]

So, from all the pyrotechnics of electromagnetism, our senses select just that narrow band of radiation that represents the visible spectrum – those wavelengths which lie between 375 and 775 billionths of a metre. We look at the world through a tiny slit, and this narrow window on reality is even further restricted by censorship taking place between the eye and the brain.

Experiments at the Massachusetts Institute of Technology have shown just how severe this restriction can be. Frogs have eyes

85

which are structurally much like our own. Theoretically, they should be able to see as well as we do, but microelectrodes implanted in frog optic nerves reveal that these pass on only very limited information to frog brains. From all the richness of the visual world, which falls largely without distortion on a frog retina, only three very basic kinds of message are relayed. Frog brains are given just a general outline of the environment, along with news about moving shadows such as those thrown by birds of prey or mammalian predators, plus a selection of more precise information about small dark edible objects such as insects flying by. The rest of the world is filtered out, discarded as of no practical importance.[226]

Our senses are less restrictive, but no less selective. When you walk into a room that contains a grandfather clock, the tick-tock seems very loud. Your brain-waves, heart-beat and skin resistance all fluctuate in time with the noisy mechanism. But after a while, you no longer find it quite so intrusive and eventually cease to hear it at all. You tune it out, your skin resistance remains undisturbed and, if measurements are taken from your auditory nerve, these show that news of the clock is no longer even being sent on from the ear to the brain.[172] The clock becomes a constant part of the environment, one without further news value. But if the clockwork rhythm changes or stops altogether, then attention returns to it immediately. Something has happened which might be significant and needs to be considered.

Neurophysiologist Karl Pribram of Stanford University describes this as the "Bowery Effect", after an elevated railroad of that name which once ran along Third Avenue in New York. It was travelled, at the same time late each night, by a particularly noisy train involved in maintenance work. People in the neighbourhood were delighted when the track was demolished, but for months after the line closed, the local police station was besieged by reports of "something strange", possibly thieves or burglars, making noises in the night. It turned out that all the calls came in at precisely the time of the former late-night train – from residents now "hearing" the absence of a once familiar sound.[287]

We have a parallel and equivalent ability to ignore such gaps in our experience altogether. All of us, for instance, suffer from a "blind spot", a permanent hole in the world left by that part of the retina which is insensitive to light because of the position of the optic tract. But we are normally totally unaware of this shortcoming and can only be convinced of it by experiments with dots on cards that reveal an ever-present gap in our perception.

Our awareness is clearly both selective and subjective. We not only tune in to restricted bits of reality, but also shuffle and sort these out in accordance with our experience and expectation of them. We arrange things into categories and patterns, sifting the limited input and making "sense" of what we do receive. We then check such understandings against the experience of others similarly occupied and finally reach a consensus about reality. The arbitrary nature of this agreement is more clearly revealed by the adjective "con-sensual", a gathering of the senses, a common response obtained by those operating under similar constraints. What we experience and describe, however, no matter how widely it is validated, is not necessarily anything like what actually exists out there in the real world.

Our consciousness of the world is biased. We see not with our eyes, but with our brains. What a piece of bread looks like, depends on how hungry we are. The same coin looms larger in the minds of poor children than it does in those from wealthy homes.[36] Children as a whole are usually more open to the true nature of things. Their experience, according to pioneer psychologist William James, is "a blooming, buzzing, confusion", which only settles into formal patterns as we mature and learn to construct a personal view which is more socially acceptable. But we should never lose sight of the fact that this perception is essentially artificial. As psychologist Robert Ornstein points out, "Nothing sacred occurs in nature between the electromagnetic wavelengths of 380 and 370 billionths of a metre, yet we can perceive one and not the other."[271]

This shaping of reality can be a highly creative process. "Nature," in the view of philosopher Alfred North Whitehead, "is a dull affair, soundless, scentless, colourless, merely the hurrying of material, endlessly, meaninglessly." It is we who give the rose its scent and the nightingale its song. "We are the music," said T. S. Eliot, "while the music lasts." As a sculptor takes responsibility for releasing one of a multitude of potential statues trapped in a block of stone, so we reveal the beauty in nature and in data, running always the risk of losing something in translation. We all have access to what Aldous Huxley called "Mind at Large", but most people, most of the time, know only what comes through the reducing valve of their own sense systems and becomes petrified in language. Reality is rife with "seas like millponds" and "nights as black as pitch". But all is not lost.

[Some people] seem to be born with a kind of bypass that circumvents the reducing valve. In others, temporary bypasses may be acquired . . . through these there flows, not indeed the perception of "everything that is happening everywhere in the universe", . . . but something more than, and above all something different from, the carefully selected, utilitarian material which our narrow individual minds regard as a complete, or at least sufficient, picture of reality.[179]

It is with such expanded views of the world, things on the brink of the supernormal, that these chapters on Mind are concerned.

Chapter Four

SELF

The most peculiar thing about biology is that it is a science named after something that cannot be defined.

For all our expertise in unravelling the genetic code and the miracles of nerve conduction, muscle movement and blood clotting, we still understand virtually nothing about pain, sleep, growth and healing. We are ignorant about nearly every aspect of consciousness and even find it hard to diagnose death. The truth is that we know life only by its symptoms. And the best definition of biology remains that of Hungarian biochemist Szent-Györgyi, who called it "the science of the improbable".[357]

Amongst all the improbability, nothing stands out more dramatically than the human brain, in the creation of which – said Arthur Koestler – "evolution has wildly overshot the mark."[214] It seems to be an instrument well in advance of our needs. Why else should it be possible for anyone to respond, just four seconds after being asked to turn 4/47 into a decimal, with the reply "Point 0851063829787234042553194." A. C. Aitken, then Professor of Mathematics at Edinburgh University, did this in just 24 seconds, then stopped to think for a moment before adding "893617021276595744468 – and that's the repeating point. It starts again then at 085."[347]

An awareness of fractions through the forty-sixth decimal place cannot, under any conceivable circumstance, be regarded as having survival value. It is hard to imagine any environmental pressure which could induce evolution to work towards such a goal. And yet it clearly has done so. Professor Aitken was, in addition, able to memorise strings of random numbers – something that is fiendishly difficult to do. The average ability is barely more than seven digits, but he succeeded on several occasions in reeling off a string of a thousand numbers without hesitation or error.[177] Others have gone even further. In 1981, Rajan Mahadevan in Mangalore put himself into the record books by reciting, from memory, the value of "pi" – the ratio of the circumference of any circle to its diameter – correctly to 31,811 decimal places.[265]

The professor and the young Indian student are not freaks. Both spent long periods practising their numerical and mnemonic skills in order to perform feats which put them at the extreme of normal human ability, but they are not qualitatively different from the rest of us. Actors routinely master enormous amounts of verbal material during their working lives and it seems likely that all our brains have the capacity for storing huge amounts of information and will do so, given the right incentive.

Before the advent of printing and general literacy, most cultural material was stored and transmitted orally and a good memory was a useful talent, perhaps even one with enough survival value for it to be favoured by natural selection. But the period during which we were advanced enough to enjoy living together, swapping tales across a campfire, and yet not advanced enough to have invented writing, was a comparatively short one in our evolution and cannot in itself account for the explosive growth of our brains. Everything we know about the evolution of the brain suggests that it was big enough and complex enough to handle feats of memory and calculation long before we found ourselves in situations where such talents might prove to be useful. Our cortical skills were undoubtedly honed and polished during the last few hundred thousand years, but it looks as though we came to the game already provided with most of the right equipment.[422]

There is an astonishing case on record of a man who was able to recall a pattern of apparently random dots with such clarity that, on being presented the following day with a second pattern, he could fuse the two in his mind. Proof that he was doing so was provided by the fact that the two pictures had been very carefully designed so that when one was superimposed precisely on the other, and only then, a square appeared in the centre.[355] This is known as visual or "eidetic" imagery and may be rare, but many of those with unusually good recall use some kind of synesthesia or sensory blending, associating numbers and names with vivid images or musical tones. [6] The mixture, it seems, lowers the threshold for all the senses and widens the reducing valve enough to expose the mind to larger concerns. This provides information and produces talents that confer no obvious selective advantage on an individual or even on a species, but they do offer fascinating glimpses of levels on which such things perhaps have reason and meaning.

We spend a good part of our time establishing personal boundaries, creating individuality by drawing lines that define the limits of the self. But it becomes increasingly clear that these limits are artificial. We are part of the fabric and cannot avoid being so. Mere anarchy is seldom loosed upon the real world.

Bioelectricity

Albert Szent-Györgyi von Nagyrapolt was conscripted into the Austrian Army during World War I. He was decorated for bravery, but deliberately wounded himself in order to get back to his studies in medicine – which led in 1937 to a Nobel Prize. He was equally cavalier about World War II, playing an active role in the underground in Hungary, but not until he had taken time out in the mad days of early 1941 to lay the foundations of a new biology.

Szent-Györgyi, which translates rather aptly as Saint George, deliberately took on the dragons of the biological establishment. When scientists break living things down to their constituent parts, he pointed out, somewhere along the line life slips through their fingers and they find themselves working with dead matter. "It looks," he said, "as if some basic fact about life is still missing, without which any real understanding is impossible." He was thinking of the mysterious force which shapes the ends of organisms, determining the destiny of growing cells, and made a suggestion which outraged the scientific world. The missing factor, said Szent-Györgyi, is probably electricity.

The reason for outrage is partly historical. In the late eighteenth century, an Austrian physician called Franz Mesmer proposed the existence of a "subtle fluid which pervades the universe, and associates all things in mutual intercourse and harmony".[183] His inspiration derived from Cabalistic and astrological tradition, and is in fact not far removed from present understandings of gravity and electromagnetic radiation, but the problem is that he became involved in an extraordinary craze. Mesmer was convinced that his fluid, acting through what he called "animal magnetism", could work miracles and cure all ills. He set up a clinic in Paris which produced public and dramatic cures amongst patients who touched his oak tubs filled with "magnetised" water. And his ideas became, not only fashionable, but also the centre of furious debate. Most scientists were appalled at the circus atmosphere in the clinic and it was not long before the French Academy of Medicine outlawed Mesmer and announced that any physician who even mentioned animal magnetism would be struck off the register. The debate nevertheless continued into the nineteenth century, with revivals in France, Germany, Sweden and Victorian England as others explored "mesmerism" and discovered that trancelike states and cures could indeed be produced in patients without the use of magnets. Hypnotism slowly became an accepted practice, but Mesmer's early association with magnetism condemned this aspect of his theory to the lunatic fringe and bedevilled all future attempts to reconcile living things with their electromagnetic environment.

A second reason for resistance to Szent-Györgyi's electric insight was, and remains, the success of biochemistry and molecular biology. The development of X-ray diffraction in 1912 and of the electron microscope in 1939, opened a new world to science – and reduced the elusive substance of life to very satisfying mechanical ingredients which could be mapped and manipulated without recourse to inconvenient and invisible forces. Early pictures of the interior of cells and molecules showed no wires, so the possibility of their conducting electricity could comfortably be ignored. And it was, and might still be, were it not for two remarkable men.

The first of these was Harold Saxton Burr, Professor of Anatomy at Yale. Burr, like others of his generation, was aware that nerve conduction was at least partly chemical in nature. The Nobel Prize for medicine in 1936 went to the discoverers of acetylcholine, an ammonium derivative which transmits impulses between the ends of nerve fibres. But Burr was convinced, like Luigi Galvani before him, of the existence of "animal electricity". His equipment in the early 1930s was rudimentary, consisting of bulky vacuum tubes and meters so "noisy" they concealed more than they revealed. But with these he nevertheless looked for, and found, small electric potentials on the surface of animals as diverse as worms, hydras and salamanders. He measured an electric charge in slime mould growing on a forest floor and discovered that a voltmeter hooked up to a tree was able to record a field that fluctuated in response to changes in light, moisture, storm activity, sunspots and the phases of the moon. Animals and plants, said Burr, "are essentially electric and show a change in voltage gradient associated with fundamental biological activity".[40]

No journal in 1935 would touch such heresy, but Burr was lucky enough to have a captive forum for his ideas. He was editor of the *Yale Journal of Biology and Medicine* and used it that year to publish his findings on the electric characteristics of living systems.[41] And continued, quite shamelessly, to make regular use of the same outlet for a further twenty-eight contentious papers outlining the bioelectric nature of menstruation, ovulation, sleep, growth, healing and disease – and to proclaim a comprehensive electrodynamic theory of life. There is, insisted Burr, a veritable "life-field" which holds the shape of an organism just as a mould determines the shape of a pie or a pudding:

> When we meet a friend we have not seen for six months, there is not one molecule in his face which was there when we last saw him. But, thanks to his controlling life-field, the new molecules have fallen into the old, familiar pattern and we can recognise his face.[40]

Beyond regular subscribers to the Yale journal, Burr's work went largely unnoticed. Where it was considered at all, it was dismissed as foggy vitalism. Biological knowledge at that time had no way to examine currents flowing within cells and no theoretical framework to explain where such electricity, or the fields it was said to produce, could come from. Burr perhaps went too far in claiming to be able to predict emotional and mental health on the basis only of a difference in voltage between a subject's head and hand, but he was at least observing and recording such a difference. Most biologists took the easy way out and ignored papers with titles like "Moon madness"[38] and "Tree potentials".[39] No one bothered to see if the measurements Burr was making were valid. He was never discredited or even accorded the courtesy of critical review. Despite twenty years of careful and methodical work, he was simply forgotten. And might well have remained so, were it not for the work of a second American iconoclast.

Robert Becker was an orthopaedic surgeon who became

obsessed by one of his speciality's unsolved problems – the occasional refusal of broken bones to grow back together again. Such failure is a disaster for the patient and a bitter defeat for the doctor who sometimes even has to amputate an unhealed limb and fit a mechanical substitute. Becker assumed that bones which wouldn't knit must be missing something that triggers and controls normal healing, and turned for the answer to animals whose bones not only mend after fracture, but can regenerate altogether. Lizards grow new tails, complete with bony skeletons, after shedding the old ones as diversions in the face of imminent predation, and salamanders – according to an eighteenth-century Italian priest – are capable of entirely regrowing severed limbs. The astonishing Lazzaro Spallanzani apparently persuaded one young salamander to replace all of its legs six times in just three months. So in 1958, Becker began his own work on these fruitful little amphibians.

Becker was encouraged by Russian evidence that regeneration in plants and animals was accompanied by a "current of injury" – an electric potential that could be measured at the site of the wound. He suspected that the electricity might be more than a symptom of the injury and decided to see whether induced changes in such currents had any effect on the healing process. Or at least that was his intention until he discovered that American science was not as open-minded as its Soviet equivalent. "The Russians are willing to follow hunches," says Becker, "their researchers are encouraged to try the most outlandish experiments, ones that our science just *knows* cannot work. Furthermore, Soviet journals publish them – even if they *do* work." The response of his research board in Syracuse, New York, was hostile. "The notion that electricity has anything to do with living things was totally discredited some time ago," they pronounced. "It has absolutely no validity. The whole idea was based on its appeal to quacks and the gullible public. We will not stand idly by and see this medical school associated with such a charlatanistic, unscientific project."[15]

Becker fortunately found another way round this political impasse and performed a vital experiment. He took a number of frogs and salamanders and, with sensitive equipment, showed that the ends of all their limbs were negatively charged with minute currents of just 0.000002 amps – that is ten million times weaker than the average household supply. Then, under anaesthetic, he amputated the right forelimb of each animal and found that the current continued to flow in the stump, but that its polarity was reversed. All now provided a positive reading. The frog wounds soon healed, producing scar tissue as their polarity slowly returned to its original condition. The salamander potentials dropped quickly at first, but then suddenly soared, turning negative and climbing within a week to three times their original level. And in this electric frenzy, they began to regenerate, developing new and completely functional limbs within a few weeks.

It took ten years to develop a tiny battery capable of imitating the salamander's natural electric instruction, but as soon as one was available and was implanted in the stump of an amputated frog, it too grew a new limb. And in 1972 a similar device succeeded in producing partial regeneration in the amputated leg of a rat, which became the first mammal ever to do so. No human has yet enjoyed similar whole-limb benefits, but thanks to Becker's insight, patients all over the world are now having small hearing-aid batteries that produce a sustained negative charge, implanted close to severe fractures or alongside those which, in older people, show a marked reluctance to heal. And the results are dramatic. Electrical osteogenesis works.[14]

Becker's strength, however, was not just as a clinical orthopaedist. He was interested in electricity as a factor in the overall control of cell differentiation and growth and went on to show that the right kind of current could – inhibit infection; relieve pain; halt osteomyelitis; restore muscle control; repair intestinal ruptures; close holes in the heart; regenerate nerve cords, perhaps even spinal cords; and replace lost parts of the brain. And perhaps most important of all, he looked more closely at the direct electric currents and fields being produced by a number of organisms and proved that they were semiconducting systems with unexpected properties.

There are three recognised ways of conducting an electric current. The first and simplest involves free electrons moving along a metallic surface, usually a wire. This is unknown in living things. The second involves the movement of charged particles or ions through a solution. This works well across living membranes, but cannot be sustained over long distances. And the third and most complex is semiconduction, which requires an ordered and regular medium and can only carry a very small current, but does so faster with increasing temperature and over extraordinary distances. Semiconduction was a laboratory curiosity during the 1930s, but has come into its own with transistors and may provide answers to many of life's persistent mysteries.

Semiconductivity occurs naturally in inorganic crystals such as silicon, changing their structure so that each time a current passes through, it does so more easily – making such crystals, in effect, very lifelike; more responsive to their environment; better able to serve one of Rupert Sheldrake's morphogenetic fields. And if the arrangement of proteins in our nerves gives us possession of such conductors, it could also give us greater awareness of subtle environmental patterns such as the fields produced by under-ground water. It could, in addition, give us access to signals on very low frequencies – the sort that travel round the globe with little attenuation – and provide information about the currents and fields produced by other organisms. It is admittedly tenuous, but all at once there is at least a prospect of coming to terms with phenomena such as dowsing, telepathy and auras.

With fifty thousand patients so far treated with electricity to mend intractable bone fractures; with countless others now being fitted with simple electric jewellery that succeeds from outside the body in stabilising heart-beat and disturbances in brain-wave rhythms that may produce epileptic fits; with coils and implants healing ulcers, severe burns and painful tendonitis – it is no surprise that electromagnetic medicine is coming of age and begin-ning to wonder what effects natural and artificial fields might have on the behaviour and physiology of whole organisms. If we are essentially electric, then what does it mean to us to live in an increasingly electric environment?

The English zoologist J. Z. Young once likened the brain to an old-fashioned telephone exchange in which operators sit in a row at a vast switchboard. They see lights flash and they plug the circuits in and out and, in the fashion of operators everywhere, they eavesdrop on the conversations. As a result of what they hear, they whisper together. And it is in this gossip, said Young, "that I discern the most intrinsic function of the organisation of the brain".[424]

It is not the traffic that we need to watch, but its consequences – the "whispering together" of the more sensitive bits of the world.

Mind Fields

Life has used electricity for ages.

Ancient fish were almost custom-built to act as detectors of stray charge. Scattered across the head and around the jaws of all sharks, skates, rays and dogfish are tiny pores, little openings in the skin which lead to jelly-filled canals. These tubules gather on the nose, below the eye and at the tip of the lower jaw into clusters of sensory cells known, after a nineteenth-century Italian anatomist, as the ampullae of Lorenzini. Since their discovery, these organs have been variously suspected of being temperature, chemical or mechanical receptors, until proof in 1962 that they are the most electrosensitive structures known anywhere in the living world.[257] Tests show that free-living skates *Raja clavata* are aware of voltage gradients as low as 0.01 microvolts – which means that they can detect a field ten thousand million times weaker than that produced by household current.[201]

All marine fish set up direct current, low frequency voltage gradients as they swim through salt water or pump it past their gills. And these fields are detectable at a distance, betraying even motionless fish to sensitive predators. A series of elegant experiments on the bottom-feeding shark *Scyliorhinus canicula* has shown that it can detect a flounder buried beneath several inches of sand at night and in muddy water, and dig up and eat the hidden flatfish in seconds. The fact that the shark is responding to electric cues was proven by enclosing flounders in blocks of agar – a gel made from seaweed, which is virtually transparent to electrical current, but effectively conceals the fish from sight, smell or sound. The sharks located and attacked flounders buried in such agar coffins just as easily as before – but showed absolutely no awareness of, or interest in, portions of highly aromatic dead flounder concealed in the same containers. It was the field produced by a living fish that attracted their attention. In the end, however, all it took to defeat the sensitivity of voracious sharks was a thin layer of polyethylene film. Once this electrical insulator was wrapped around a live flounder, it became virtually sharkproof. A discovery which has not been wasted on the United States Navy, which is now testing a personal shark screen that fits into a pocket and consists of nothing more than a simple body-sized polyvinyl bag suspended from an inflatable collar.[202]

In 1831, at one of his vastly popular public lectures to the Royal Institution in London, Michael Faraday took a copper wheel and turned it so that its rim passed between the poles of a permanent magnet.[106] An electric current was immediately set up in the disc, led off and put to work, thus making the greatest single electrical discovery in history – the invention of the first electric generator. Or at least the first human version. Faraday was in fact pre-empted by several hundred million years.

Earth is a permanent magnet with its field stretched between north and south magnetic poles. When a conductive body like a shark cuts across this field by swimming west to east, it acts in the same way as a copper wheel, inducing an electric current that flows through the fish, setting up a subtle field appreciable to the ampullae of Lorenzini. In other words, a shark can by its own motion, acquire an electromagnetic compass wherever it may be. And learn to orient itself, not only with respect to this handy personal probe, but also to much larger and more stable fields such as those induced by ocean currents that interfere with Earth's field in the same way. The fact that many animals do precisely this, is only just beginning to be appreciated.

In 1975, Richard Blakemore of the University of Massachusetts made the astonishing announcement that bacteria had a magnetic sense. While studying the salt marshes of Cape Cod, he noticed that simple rod-like microbes there always arranged themselves facing northwards on his microscope slides. He became convinced that the bacteria were using the declination of the Earth's field as a guide, against the random molecular motion of the water around them, down to the rich bottom mud. And confirmation of this theory soon arrived in reports of bacteria and algae at Rio de Janeiro and in New Zealand which migrated southwards along the same magnetic lines down to their equally attractive southern hemispheric oozes.[414]

Electron micrographs show that all these microbes contain magnetite microcrystals, each particle a single domain, the smallest piece of the mineral that can still act as a magnet. Similar tiny built-in lodestones have since been found in the stomachs of bees, and in 1984 a group headed by Michael Walker of the University of Hawaii isolated magnetite deposits in the skull bones of migratory fish such as yellowfin tuna *Thunnus albacores* and chinook salmon *Oncorhynchus tsawytscha*.

Pigeons have a minute deposit of magnetic ferritin, a protein capable of storing iron, on the right side of their heads, between the brain and the inner surface of the skull. And at the University of Lund in Sweden it has been nicely demonstrated that, although birds are born with this natural compass, they have to learn to calibrate it for themselves. Pied flycatchers *Ficedula hypoleuca* – plump little birds with white wing patches, that breed in northern Europe but winter in West Africa – were reared in nest boxes fitted with Helmholtz coils. For the first 12 days of their lives, the fledgling birds lived in magnetic fields rotated through 90 degrees so that magnetic north was shifted to the east. When autumn came, they were living under the normal magnetic influence with a clear view of the sky and access to sun, stars and polarised light, but they ignored these common celestial clues and chose instead to head off at right angles to their traditional migration route. Had they not been restrained, the hapless young flycatchers, dominated by their early magnetic brainwashing, would have ended up somewhere in the frozen north Atlantic.[1] And the possibility exists that we are becoming similarly confused.

In 1961, Robert Becker began to look at the effects on humans of disturbances in Earth's field caused by magnetic storms in the sun. He compared the admission of over 28,000 patients to psychiatric hospitals with sixty-seven severe magnetic storms recorded during the previous four years, and found a strong correlation. Significantly more people were admitted just after magnetic upheavals. He watched schizophrenics already in hospital and found marked changes in behaviour just when low energy cosmic ray flares were disrupting Earth's field. He found high levels of stress, abnormal endocrine activity and slowed reaction times during sunspot activity. And concluded that our body's direct current electrical control system is tuned by natural rhythms and is responsive to changes that take place at surprisingly low levels. A conclusion strongly supported by recent work on biological cycles, particularly that of Rutger Wever in West Germany, who has found that people kept from all contact with Earth's field in an artificially shielded environment, become thoroughly desynchronised.[404]

For the first 4000 million years of Earth's history, our planet's electromagnetic environment was fairly simple. There was a weak background field modulated by terrestrial pulsations and sculpted by lunar and solar cycles. There were bursts of static from lightning and a few weak radio waves from the Sun and other stars. There was light, of course, and a small amount of high frequency radiation from space. But large parts of the energy spectrum were totally silent. The world was quiet in a way we will never know again.

During the last forty years, everything has changed. The technological boom which began after World War II has left us awash in a sea of strange energies. Every digital watch, flashlight and portable radio produces a direct current magnetic field. The stopping and starting of each electric train turns the power rail into a gigantic antenna that radiates low frequency waves for over a hundred miles. High voltage power lines, of which there are half a million miles in the United States alone, produce powerful fields that are concentrated by metal objects nearby, and pass through switching stations that also emit radio frequencies. Low frequency radio waves emanate from navigation beacons and military networks. The medium frequencies are loaded with AM transmitters, and the high and very high frequency channels bristle with police and taxi traffic, the sibilance of spy satellites and the ceaseless babel of millions of CB radios. And in addition, we are bombarded at home and at work, every hour of the day and night, by errant microwaves from antitheft devices, metal detectors, automatic door openers, walkie-talkies, cordless telephones, vulcanisers, heat sealers and ovens.[15]

We are, without permission but with our tacit approval, the subjects of a giant electrical experiment. Nor is there any end in sight. The density of radio waves around us now is 100 million times the natural level reaching us from the Sun, and by 1990 it will have doubled again. When superconducting cables are introduced, the field strength around power lines will be increased by another twenty times. And electric cars and vehicles moved by magnetic levitation will add entirely new sources of electropollution to the stew with which we are already assailed. Meanwhile, the first results of the experiment are starting to come in and there is, it seems, no place to hide.

Since 1982, a high incidence of leukaemia has been reported in three independent studies of people exposed to high electromagnetic frequencies – those who work as radio operators, electronic technicians, power linemen, aluminium smelters and electrical engineers in California, England and Wales. An increased susceptibility to cancer has been found in children whose homes lie near high voltage power lines. Helicopter pilots, who may be exposed to more radar than the rest of us, are producing an unusually large number of club-footed children. Pregnant women operating video display terminals in the United States, Canada and England appear to suffer miscarriages, have stillbirths or give birth to deformed children far more often than the population as a whole. And large numbers of people in highly electrical environments everywhere complain of headaches, loss of appetite and frequent fatigue. The trickle of such reports grows to a flood, provoking some consumer resistance and predictable disclaimers from the power companies and other vested interests. But perhaps the most disconcerting aspect of electromagnetic smog is that those, like Robert Becker, who know most about the biological effects, are least concerned about the more obvious sources of contamination. It is the low-intensity radiations that worry them – the ones closest in strength to Earth's own subtle signals. These are the ones, according to a Soviet scientist, that "alter visual, acoustic and tactile sensations, changing physiological functions and affecting emotional states".[259]

Russian thinking about bioelectricity is conditioned by Pavlovian psychology and an awareness that animals can respond to very subtle signals. This knowledge and an early interest in biometeorology, has shaped their theories about the function of the brain and led to the establishment of far lower tolerance levels and far higher, and probably more realistic, national safety standards in the manufacture and operation of electrical machinery. It may also have led to some hitherto rather puzzling international applications. Why else would anyone go to the trouble, over a period of thirty years, of beaming a weak and apparently ineffectual pattern of microwaves at the American embassy in Moscow? Unless such a stimulus can, or it is believed that it can, modify brain-waves and behaviour.

In 1976, the government of the United States ordered aluminium shielding to be nailed to the sides of their embassy building and the State Department initiated a series of tests in which primates were exposed to microwaves similar to the Moscow signal. The shielding, together with a 20 per cent danger allowance, went some way towards appeasing embassy staff, but the outcome of the experiment is still unclear. Although it is said that one of the monkeys involved fell into a deep stupor after ten days' exposure to the simulated microwaves, and could not be roused until the power was cut off three days later.

I have dealt elsewhere with the strong evidence for environmental control over many basic biological systems.[396] Our internal clocks are clearly tied to the rhythms provided by our planet and its nearest neighbours in the solar system. We wake and sleep, sweat and shiver, urinate and respirate in time with cosmic cues that are often so subtle that medical science has had a hard time taking them seriously. But an avalanche of studies during the last decade on insomnia, menstrual irregularity and stress in those suffering from cyclic disturbances such as jet lag, has turned the tide. It is now more widely accepted that functional integrity, the basic processes of growth and control, and the efficient working of the central nervous system are all maintained to a very large extent by our electromagnetic environment.[250]

The extent of this influence is being calibrated by laboratories such as that run by neuroscientist Ross Adey at Loma Linda Hospital in California. He and his colleagues have found that monkey brain-waves rapidly become entrained to rhythms of conflicting fields in their environment, and that these can produce marked changes in behaviour. He has succeeded in altering response time and affecting short term memory simply by exposing monkeys to low frequency fields no stronger than those produced by a colour television set at 60 feet or by a light bulb burning 10 feet away. And he suspects that an interaction between radiation and body chemistry might make us even more susceptible to fields too weak in themselves to trigger a nerve impulse.[259]

We have all the clues we need to make the link between microwaves and sensitive human central nervous systems. "Confusion beam" weapons are a theoretical possibility and the Russians may already have found an effective window through which to point them, but we don't need to go to Moscow for evidence of such effects in action. In 1979, Stephen Perry, a physician working near Wolverhampton, noticed that people in his practice who lived near high-voltage lines seemed more prone to depression. Reasoning that suicide was an overt and unequivocal sign of such depression, he plotted the addresses of 598 suicides on maps showing the location of power lines in his area. And then compared this distribution with a set of another 598 addresses chosen at random. Houses in which suicides occurred were, on average, closer to the wires. Perry confirmed this link by going out and measuring the actual field strength and discovered that it was 22 per cent higher at suicide addresses, and that those areas with the strongest magnetic fields contained 40 per cent more fatal locations than randomly selected houses.[237]

The conclusion seems as unequivocal as the fact that four unrelated people living in houses that nestle under one pylon in Walsall, have killed themselves since 1971. The odds against that happening purely by chance are over 5000 to 1. We are, it seems, electric animals with sensitive magnetic minds, caught up in a web of electromagnetic influence. And it is somewhere in this energetic web that we have to look for the impetus that made us conscious and gave us the ability to knit ourselves into such a tangle.

Biofeedback

Venus is the brightest of planets. It shines with light reflected from the Sun and passes through distinct phases from crescent to full, just like the moon. As a "morning star", it glows with a magnitude of − 4, making it almost 200 times as brilliant as the true stars of the Plough. And as the sun rises and the sky lightens, Venus remains just as bright as ever, but it becomes relatively dim and disappears from view. At least it does from my view.

Sometimes, from the bottom of a mine shaft, or shielded from the sun in the deep canyons of a great city street, I can look up still and see our sister planet high in the sky at noon. There are, however, those who can do this at any time, anywhere, pointing out the planet with unerring precision as they go about their tribal business in the Andes or the Kalahari. A few hundred years ago, sailors from our own civilisation navigated with the aid of Venus, following its path as easily by day as they did by night; but something has happened to us in the interim. Venus is still there, passing through its customary phases, transitting the sun, but we have changed. We have chosen to develop a different set of skills. We make precision tools and travel to the moon, but we have lost sight of Venus.[77]

Something of the same sort happened during the evolution of the nervous system.

The first nerves, in worms and other early invertebrates, were simple telegraphs, coordinating the behaviour of collections of relatively independent cells. Later, amongst insects and molluscs, these lines of communication grew into more elaborate networks, expanding to form centres of far more careful control. And eventually, with the development of the vertebrates, bodies became so intricate and their relationships so complex, that it was necessary to have separate systems for dealing with different environments. Responsibility was split between internal and external affairs, and awareness came to be similarly subdivided.

Ordinary consciousness is now largely concerned with our external environment, making us almost totally unaware of what goes on inside. It has to be this way. We could not survive if we had to make our hearts beat voluntarily seventy times each minute or think about the need for every breath. There would be no time for anything else if we needed to mastermind the process of digestion, measure out hormones from our glands, or direct the operation of the kidneys. So all these functions have been taken out of direct awareness and put on automatic. They have been placed under the unconscious control of the autonomic nervous system.

The nerves that direct skeletal muscle, the ones concerned with voluntary action and manipulation of the world about us, are identified, along with the brain, as the central nervous system. This division of labour between voluntary and involuntary nerves is real and necessary. Our busy lives would be impossible without it. But the split is now so deeply enshrined in current scientific thought that we tend to see the two systems as not only functionally, but biologically distinct. We have, with our labels, effectively legislated the autonomic system beyond self-control and made it almost impossible to bridge the gap. And have come, as a consequence, to regard anyone who claims to be able to do so – who cuts into skin without bleeding, or is buried alive or walks through fire – as a charlatan of no possible interest to science.

An adept of yoga, someone with a different mental set, from a tradition that encourages self-regulation, has fewer problems. One particular yogi, under examination over a decade ago, showed that he was able to change the temperature of two patches of skin on his hand, making one hotter and the other simultaneously colder. The two areas were only a couple of inches apart, but showed a temperature difference of 10°F, the hot area glowing red and the cold patch turning an ashen grey.[314] This demonstration failed at the time to excite clinical physiology and was largely ignored as evidence, but attitudes have changed in recent years and it may not be too long before we begin to learn how to see Venus again.

The tide began to turn in 1924 with a series of experiments in the Soviet Union by students of Ivan Pavlov. These showed that dogs could be conditioned, not only to salivate on cue, but to change their body heat – even to the extent of controlling blood flow to one leg at a time. "The gap between the two disconnected worlds of psyche and soma," the workers concluded, "is being bridged."[42] It was narrowed even further during the 1960s by Neal Miller and his colleagues at the Rockefeller University in New York.

Miller worked with rats and instead of offering his animals the usual incentives of food and drink, rewarded each successful performance directly with a small electric shock to the "pleasure" centre of the brain. To obtain this "buzz", the rats proved both willing and able to salivate, produce extra urine, speed up and slow down their hearts, alter blood pressure, control the circulation in their stomach walls, or direct warmth to one ear rather than the other. In one extraordinary experiment, a rat even succeeded in learning how to fire just one single nerve cell at a time. And they did all this on demand and at will, by exercising their will at the behest of the scientists.[246] On being asked about the relevance of this work to human physiology and behaviour, Miller wryly observed, "I believe that in this respect, humans are as smart as rats . . . but this has not yet been completely proved."[110]

Work is in progress.

At the University of Pittsburgh, students have shown over a period of weeks that they can acquire control of their heart-beats and keep these going within very narrow limits.[220] Other students at Harvard claim to have picked up the trick in just half an hour.[337] And in Kansas a yogi called Swami Rama has succeeded in stopping the flow of blood to his heart altogether.[33] In Baltimore, these autonomic callisthenics have been put to good use by training heart patients to overcome the rhythmic disabilities of atrial fibrillation and premature ventricular contraction.[104] And at the Lafayette Clinic in Detroit, at Harvard Medical School and at Rockefeller University, subjects suffering from high blood pressure have learned to bring this down and keep it down without the use of drugs.[95]

Students at Oxford University have been trained to vary the

temperature in their earlobes, up or down, one at a time or both together.[352] While at the Menninger Foundation, housewives have learned to vary their hand temperature as effectively as trained yogis.[136] Some have discovered, as a byproduct of this ability, that drawing blood away from the head to the hands is a sure cure for the sharp pains of migraine.[135] The dull discomfort of tension headaches, on the other hand, has been dealt with at the University of Colorado by simply teaching sufferers how to relax the frontalis muscle in their foreheads.[37] The salivary reflex has been successfully brought under conscious control at the University of Queensland in Australia.[407] And, just to show that anything that rats can do . . . at Queens University in Canada, human subjects have been trained to discharge a single motor-nerve cell at a time – just one selected from the brain's array of around ten billion.[10]

And all this wizardry is being accomplished by a simple technique. Abilities once regarded as paranormal are being made available to normal control just by letting the conscious mind know what is going on at unconscious levels. In the esoteric tradition of yoga, this is achieved by tuning out the noise of competing activity, by turning day into night so that Venus becomes visible. But for less disciplined Western minds, it has proved a great deal easier to amplify the signals themselves, making Venus bright enough to see even in the daylight brilliance of active consciousness.

The technique is called biofeedback. It is a way of learning, or rather revealing, what the body already knows by hooking internal processes up to an external display so that you can see or hear what your organs are doing. That is all, and it is usually enough to bring the processes under conscious control.

The most useful biofeedback instruments include an electrocardiogram, which monitors heart-beat; an electromygraph, which measures impulses associated with muscle tension; a skin resistance meter, which gives a direct measurement of arousal; a thermistor, for taking the temperature of local areas of skin; and an electroencephalograph, which monitors activity originating in the brain. An ability to regulate heart, muscle and skin response has proved to be clinically useful, but it is the control of brain-waves that is turning out to be most fascinating and seminal.

The human brain hums along like "an enchanted loom, where millions of flashing shuttles weave a dissolving pattern", but it does so within a narrow range of frequencies.[343] A sleeping brain produces voltages that peak between 1 and 4 times a second (the delta rhythm); while dreaming or in a drowsy or meditative state, the shuttle accelerates from 4 to 7 cycles per second (the theta rhythm); in a condition of "relaxed awareness", a sort of neutral position, it fluctuates between frequencies of 8 and 13 (the alpha rhythm); and in a normal active or calculating state, the loom usually operates in the range from 13 to 30 cycles a second (the beta rhythm). Under ordinary circumstances, we remain unaware of these rhythms. But if electrodes are placed on the scalp at several points, and sensitive equipment is used to detect, amplify and announce the weak internal signals, we become conscious of them and enter into a new and more intimate relationship with ourselves.

One consequence of this awareness is an ability to control the brain-waves almost at will, concentrating on the production of one rhythm rather than another. And as sustained alpha rhythms in the frontal and central regions of the brain are known to be character- istic of Zen, Yoga and Sufic adepts in a state of deep meditation, numbers of people in the West began to look at biofeedback techniques as short cuts to mystic enlightenment. Electric Zen became the rage, a new and fashionable parlour game, in the 1970s and the market was flooded with devices for inducing instant alpha rhythms.

The most astonishing thing about these devices is that many of them actually work. Even the simplest can make it possible for anyone to know when alpha or theta rhythms are present, which knowledge helps to sustain such brain activity, which in turn helps people to relax. None of this, of course, is a guarantee of any kind of enlightenment. Eastern adepts use meditation as a means and not as an end in itself, but the craze has at least produced a shift in emphasis in large parts of our culture. A move away from indi- vidual, linear and analytic processes; towards more communal, holistic and intuitive ways of seeing the world.

The popularity of simple, inexpensive biofeedback equipment has focused attention on the fact that, in addition to the three recognised states of awareness – those of waking, sleeping and dreaming – there seems to be a fourth one. Deep relaxation, whether it is induced by alphameters or by easily mastered tech- niques such as transcendental meditation, is a state with clearly defined physiological correlates. Anyone in this condition shows, in addition to slow alpha and theta waves – a decrease in oxygen consumption; a reduction in carbon dioxide elimination; reductions in heart rate, blood pressure, blood lactate, blood cortisone and muscle tone; and increases in finger temperature, the perfusion of internal organs and basal skin resistance.[392] There is every reason to accept that someone in this "alert hypometabolic" condition, is demonstrating a fourth major state of consciousness.[391]

Being in such a state seems to be beneficial. Anxiety, strain and tension decrease and there appear to be improvements in general health and energy. I find that a little of it at regular intervals helps to avoid stress and acts as a sort of psychic lubricant. Others, more dedicated or perhaps just more patient than I, claim that it can even lead to a different kind of understanding, a more direct perception of reality.

As a young man, the English chemist Humphry Davy experimented with the therapeutic properties of various gases. He took extraordinary risks, inhaling everything that came to hand, finally hitting the jackpot with the discovery in 1800 of nitrous oxide. He found its effects exhilarating. "I existed in a world of newly connected and newly modified ideas. I theorised. I imagined. I made discoveries." Laughing gas quickly became the LSD of its day and was sniffed at special parties; once by Peter Mark Roget, the author of the thesaurus. "My ideas succeeded one another with extreme rapidity, thoughts rushed like a torrent through my mind, as if their velocity had been suddenly accelerated by the bursting of a barrier."[110]

It is no accident that nitrous oxide turned out to be the first chemical anaesthetic. When used in concentration, it alters brainwaves sufficiently to produce unconsciousness, but in lower doses it simply distorts the usual rhythm, releasing inhibitions and inducing a childlike playfulness which is one of the hallmarks of creativity. Consensus is rare in psychology, but most workers in the field agree that creative thinkers can be recognised by their ability to entertain wild ideas without feeling the usual need to pass judgement on them. They ask questions unceasingly and are never satisfied with easy answers. They are eclectic, never afraid of making mistakes or going beyond what is known. Their sensory perception is keen and everything becomes a source of inspiration. They behave, in other words, like children or people under the influence of laughing gas – or like those who can linger long enough in the fourth state of hypometabolic alertness.

Psychedelic agents such as LSD are not strictly comparable. They undoubtedly produce altered states, but result in brain-wave patterns that are far more like those associated with bizarre dreams. Those under their influence hallucinate, perceiving things in the absence of appropriate stimuli, whereas meditation seems to supplement and enhance perception of what is actually there. The fourth state not only relaxes the body, but widens the reducing valve of the senses and lifts the censors of the mind. It lets us, who physicist Raynor Johnson likens to prisoners condemned to looking at the world through five narrow slits in a tower, get a real look at the landscape through a gaping hole in the roof.[196]

If this clarity is sustained, it can even produce what Saint Paul called "the peace that passeth understanding". Which seems to be the same thing that Zen Buddhists know as *satori* or *kensho*, that Yogis describe as *samadhi* or *moksha*, that Sufis speak of as *fana*, and that in Taiwan they call the "Absolute Tao". All these experiences are literally ecstatic, they lift those involved into new and more spiritual realms, giving them in the process sensations of dazzling light and charismatic changes in personality. But all seem in essence, no matter how they may have been induced, to be the result of disinhibition. They are produced, not so much by adding some new and startling ingredient to consciousness, but by the removal of a barrier to true awareness.[43]

The nearest equivalent in basic biology would, I suppose, be something like the sudden startling ability of a male praying mantis who, in some species, doesn't seem to be able to copulate successfully with a female until she very effectively disinhibits him – by biting off his head.

We don't need to go quite that far.

Transcendence

There is a spur of weathered sandstone at the southern tip of Africa. It ends in a steep slope of scattered heath and protea, the last resort of a troop of displaced baboon, and the site of a lighthouse. The west wall of this defiant building is drenched by the Atlantic Ocean, by spray whipped off the surface of the cold Benguela Current that sweeps up from the Antarctic. The east wall, just a few feet away, faces the calmer waters of the Indian Ocean, touched by a flick in the tail of the warm Agulhas Current that carries the scent of coral all the way down from the tropics. This clash of two great ocean rivers is frequently concealed in dense fog, a direct consequence of their disparate characters and the reason for the light and its horn at the end of the world.

The disturbance and the danger are real, but the division between Atlantic and Indian oceans is irrelevant and arbitrary. They are human artefacts, manmade labels that have no meaning and no effect on the continuity of habitat in the west wind drift. We can no more stop the oceans from mixing than we seem to be able to separate mind and matter, mental and physical events.

It is clear that biofeedback techniques, in particular those which involve learning control of brain-wave patterns, can produce psychological changes and altered states of consciousness. And it is equally clear that meditation techniques, which are aimed at shifts in consciousness, can alter brain-wave activity. The condition of the brain and the phenomena of consciousness appear to be as inseparable as the waters of the one great world ocean. But we have at least begun to plumb the depths of that internal ocean and to appreciate that it holds some surprises. And the largest of these, perhaps, is the realisation that knowledge is not exclusively rational.

Information comes to us sometimes in a flash, in no more time than it takes to draw a breath, to have an inspiration. It seems to have an energy of its own, something Sufis describe as *baraka* and Jewish tradition identifies as *baruch*, a "blessing". It can be triggered by meditation, deep prayer, fasting, psychedelic drugs or even the onset of acute psychosis. But at its best it is spontaneous. It just arrives out of the blue, sliding into consciousness when one least expects it.

The effects on people everywhere are very much the same. They describe "radiant light", "surges of power", "eternal energy", "boundless being" and "perfect unity". There is a loss of ego boundaries, a sudden identification with all of life, a feeling of liberation and a loss of fear, a blending of the senses, bringing sharp awareness of pattern, and a sense of passing beyond space and time.[149]

One report describes how a simple flower border "became a thousand times more brilliant . . . I was not only seeing the colours

– I was hearing the colours! Each was an indescribably exquisite musical sound, the whole making a harmony that no instruments could produce."[197] Another found a glass of wine in the sunlight transformed so that it "completely engulfed me and invaded all my senses. It was a moment of pure breathtaking, absorbing ecstasy . . . I emptied my heart in gratitude."[338]

The biologist Sir Alister Hardy believed that such experience was not only widespread, but rooted in our natural history. In 1969 he founded the Religious Experience Research Unit at Oxford University, where over 4000 individual case histories have now been collected. These transcend age, sex, education, geography, cultural and religious backgrounds, and point very strongly to a common core. A conclusion which is supported by similar surveys carried out by the National Opinion Research Centre at the University of Chicago and the Religion Research Centre at Princeton University. All agree that a large proportion of people all over the world undergo a shift in awareness that is profound and often identified as sacred, but is not necessarily connected with traditional religion. It seems to have more to do with our physiology than with our cultural expectations.[159]

Psychiatrist Eugene d'Aquili of Pennsylvania Medical School has spent years on the track of a neurobiology of such transcendence. He suspects that it has something to do with the split of the brain into left and right hemispheres, each of which appears to have its own set of preferred functions. A great deal has been written recently about our "half brains", some of it far too partisan, but the fact remains that the halves do tend to specialise. Broadly speaking, the left hemisphere is primarily responsible for analytic and logical functions, while the right hemisphere controls artistic and more musical abilities. Under normal circumstances, information passes between the two hemispheres across a thick band of nerve fibres called the corpus callosum. But under certain conditions, messages get rerouted through the lower lobe of the right brain, through an area known as the "limbic system".

This part of the brain is sometimes called the nose or smell brain. It clearly has something to do with olfaction, but also seems to play a large part in the control of powerful emotions. It is a mammalian invention and the seat of the "pleasure centre" that keeps wired-up rats and cats pressing self-stimulating switches as often as ten thousand times an hour. But it also appears to lie directly between autonomic areas and the bulk of the modern conscious brain. It is a link between voluntary and involuntary nervous centres.

Each half of the brain speaks its own language. Most messages from the right hemisphere are broken down by the left into its own verbal, analytic vocabulary. The right relies on the left for speech, so any message received by the left has to be translated and emerges only as an approximation of the original despatched by the right. "For example," says d'Aquili, "a thought like 'Look at that gorgeous sunset. There must be a God', is far too vague and metaphysical for the left brain and would never make it to the conscious mind. Instead you would comment off-handedly on the sunset's colours. But when the limbic system gets involved, the thought travels from right to left brain uncensored, because emotions drive it home."[23] In other words, a mystic perception can be made convincing to the left brain – as long as it is accompanied by a powerful and simultaneous surge of emotion. And the result then is a transcendent experience.

Support for this impartial theory is provided by the fact that such experiences are not invariably uplifting. Some exposures to "boundless being" can be very negative. A few who get a glimpse of wider reality find it extraordinarily depressing. Goethe, himself a product of eighteenth-century German philosophic gloom, imbues one of his heroes with a terrible sense of despair. "My full, warm enjoyment of all living things that used to overwhelm me with so much delight and transform the world around me into a paradise, has been turned into unbearable torment, a demon who pursues me wherever I go . . . I reel with dread. I can see nothing but an eternally devouring, eternally regurgitating monster."[126] A monster that eventually consumes Young Werther, who blows his brains out in an act that was emulated at the time of publication by a number of real young people who, similarly afflicted with *Weltschmerz*, followed him into hopeless suicide.

Every age has its Cassandras, but by and large we seem to be triggered by electrifying emotion into a state of exhilaration. Into what psychologist Abraham Maslow has described as "peak experiences", which leave us exultant.[238] The free-thinking physician Andrew Weil goes a step further. He suggests that stimulation of this kind is essential and that unless the limbic system is occasionally brought into play, pushing us beyond normal sensation, giving us a glimpse of the world without the filters in place, we will suffer. "If we never learn to open the channels by disengaging our minds from ordinary consciousness, we condemn ourselves to sickness."[401]

We do seem to possess an innate drive to transcend. Three-year-old children in every culture love to whirl like dervishes until they experience vertigo and collapse. We appear to be programmed to seek out nonordinary awareness, perhaps even as a way of achieving emotional equilibrium. And it is not hard to see how such a condition could have evolved as one of the consequences of being conscious. Each stage in evolution, each addition to the size and complexity of the brain, has enhanced our awareness – producing, in effect, a new and altered state. But as the brain grew, acquiring new functional areas, we kept hold of the old ones, tacking them on at the back. And we remain, to a very large extent, under the less mindful control of these primitive relics.

It is as well to be reminded of how tight such control can be.

The broadleaf forests of eastern North America swarm during abundant summer months with a variety of small birds that come there to breed. Striking amongst these seasonal visitors are iridescent indigo buntings that decorate wood margins and hedgerows with sparks of neon. Nesting is accomplished by August and in late September the adults gather into huge flocks and leave for their traditional winter resorts in Brazil and Argentina.

Some of the season's fledglings take off on this mass migration, but a number are left behind to find their own way out in early October. These novices nevertheless do very well, moving steadily south, travelling only by night and resting up in suitable woods and shelters during the day. This progression was once believed to give young birds better protection from the predators of the day, and it might still serve such a function, but it has now been discovered that buntings fly by night mainly because they navigate with the aid of carefully selected stars. And tests in planetaria have shown that indigo bunting chicks have an innate predisposition to learn and respond directly to one particular guiding light – the North Star. They do this quickly and automatically and not only ignore the confusion of other star patterns, but are actively inhibited from reacting to these in any way.[103]

Baby buntings are "prepared" to learn this special stimulus and barred from learning others. They come into the world with a programme already largely written for them. Herring gulls are similarly circumscribed. They quickly learn to distinguish their own newly hatched chicks from the thousands of others in a colony, but never learn to recognise their own eggs – which are nevertheless equally distinctively marked. Eggs, of course, are immobile and less likely to wander off and cause confusion, so a parsimonious nature has made economies where it can, not bothering to give some species talents they are unlikely to need. With others it has been less circumspect, although there is considerable evidence to show that even we operate under severe constraints.

It is part of human conceit to think that, given enough time and effort, we can learn or be anything. But there are limits to what can be mastered even by a genius, and it is beyond dispute that all of us pick up certain skills more easily than others. As children, we pass through clearly defined stages, growing step by step from reflex-dominated infants into egocentric and then finally sociable individuals. The formation of emotional bonds, the growth of moral codes, and the acquisition of language, all follow timetables that are too short and too precise to be the result of random learning. The mind is clearly not a blank slate on which experience draws unique and intricate pictures. It is, in the words of zoologist Edward Wilson, "an alert scanner of the environment that approaches certain kinds of choices and not others in the first place, then innately leans toward one option as opposed to others and urges the body into action."[418] There is sufficient flexibility in this response to distinguish one individual from another, but the rules are still rigid enough to produce an overlap, a pattern of convergence we identify as human nature.

Our programming might not be as tight as that of gulls and buntings, but we nonetheless come equipped with instructions that have to be followed quickly and automatically to have survival value. Some orders are almost immune to logic or reason. Phobias, for instance, are so deeply rooted that they persist, despite the fact that they are obviously antiquated. Most of us now live in urban environments, but we still respond with fear and avoidance to snakes and spiders, when it would be far more useful to be instinctively wary of electrical outlets or sources of harmful radiation.

As a rule, most species come to live in habitats or niches with a limited range of resources that require a limited response. They adopt stable strategies to deal with relatively stable surroundings. Humans, on the other hand, have been pushed by rapid cultural change into a variety of habitats where traditional response systems cannot cope. We have been forced to adapt, to make use of a new talent that was lurking in the wings of evolution. We have had to learn to make up our own minds.

Once upon a time, we had ancestors that were not conscious. They had good brains, but blank minds. They were capable of receiving a variety of information and acting upon it in accordance with set programmes, as indigo buntings still do. There is nothing wrong with such behaviour. Relatively mindless buntings seem to suffer no great handicaps in life. They eat and mate, sleep and migrate on cue and keep the woods of Carolina supplied with sufficient new buntings to decorate each fleeting season. But somewhere in the line of human evolution, the pattern changed. We became self-aware and began to ask awkward questions and to abandon some of the old programmes.

The change seems to have taken place somewhere on the African savannah when our ancestors left their gypsy ways and sought a new life as hunter-gatherers. "They sought it with stone tools, they sought it with fire; they pursued it with forks and hope. But above all," says Nicholas Humphrey, "they sought it through the company of others of their kind."[175] And ran headlong into trouble.

Community life brought the benefits of an exchange of materials and ideas, but it also made our ancestors subject to all the loves, hates, spites and charities of human society. It was in this first of many schools for scandal that we learned to see ourselves and others as individuals, as free agents – rational, mindful, conscious beings with a degree of control over our own destinies. And this heady freedom had extraordinary results, putting us, amongst other things, on the surface of the moon. But graduation from those schools has also given us pause for thought and new awareness that the freedom we enjoy is qualified. We depend as much as we ever did on the rest of our environment.

No life, no mind, no thought or inspiration can exist in isolation. On its own, a mind is deprived of the nourishment it needs – a fact which may account for the vacuous nature of most alleged spirit communications. But given root in nature, there seems to be little that the combination of mind and matter, brain and body cannot accomplish. And given half a chance, it seems predestined to go a step beyond ordinary consciousness with its petty restrictions and into supersensory space.

All it takes is a little help from our friends.

Chapter Five

SOCIETY

When J. B. S. Haldane was asked by a fellow humanist whether he was prepared to lay down his life for his brother, the great geneticist replied, "No. But I would do it for two brothers or, failing that, eight cousins."

The mathematics of altruism are now familiar to anyone touched by the controversy over sociobiology. Family ties and kinship have been critically assessed from the gene's point of view, making a kind of chemical sense of self-sacrifice and nepotism. The arithmetic may be depressing, reducing potentially noble gestures to questions of cost and benefit, but it works. It is obvious where the advantage lies, and it is becoming equally obvious that the rules of kin selection apply even to single cells.

The human body is composed of about 1000 billion cells, all working towards a common end – the maintenance of a healthy and successful individual. The cells in skin, brain, blood or kidneys are all products of the same fertilised egg and share the same set of genes. Their mass may have multiplied 5000 million times, but they keep a selfish interest in the body as a whole – which may be why they usually get along so well. It is only when cells mutate, as they seem to do in cases of cancer; or drift a little from the norm, as they appear to do in old age; that they become less concerned with the fate of the community and less responsive to its needs.[8]

In the interests of harmony, it is vital that the cells in a body are not only genetical twins, but capable of recognising each other as such. Any failure to do so results in auto-immune diseases, which range from allergies to AIDS. How cells succeed in making such complex assessments remains largely mysterious, but may have something to do with "supergenes", telltale molecules scattered over their surface like familiar faces or reassuring old school ties.[74]

There is, of course, one common and recurring condition in which human bodies tolerate a gross intrusion of alien genes without vigorous resistance. A foetus is the combined product of half a mother's normal gene count with half of the father's complement. It is as foreign to the body as a skin graft and ought to be rejected. All the machinery for normal self-defence is there. The placenta has the same ability as any other organ to repel intruders; the foetus is not completely isolated from the mother; and there is ample evidence to show that pregnant women are not in a state of immunological inertia. Pregnancy ought to be impossible. The fact that it goes on happening anyway is another of life's major mysteries – one which implies that there is something more involved than the simple selfish interests of the gene.

There seems, in addition, to be a basic recognition of social benefits – of the safety and the strength, of new evolutionary vigour, to be found in numbers.

Connections

Genes stick together when they can. The closest societies tend to be those whose individuals share the greatest number of genes. And many begin to exercise such preferences even before they are born.

Experiments with rats show that blind, hairless embryos learn to recognise their siblings from clues carried by the amniotic fluid in which they float. Even if removed by caesarean section and isolated, young rats show a clear ability to recognise, and a preference for socialising with, their siblings.[163] The fact that this tendency is not innate and that learning is actually taking place in the uterus, has been demonstrated by injecting a little apple juice into the embryonic pool. Rats exposed in this way to such a novel stimulus, show a marked preference for it later, choosing to drink apple juice when other unconditioned rats opt for water.[354]

Bruce Waldman of Cornell University has shown how similar preferences enrich the family life of tadpoles which, on the face of it, would seem somewhat limited. By the time that most tadpoles struggle out of their gelatinous egg mass, their parents are long gone, but this does not necessarily deprive them of fellow feeling.

Waldman took all the tadpoles from one clutch of the toad *Bufo americanus* and dyed them red. Another clutch he coloured an equally patriotic blue. When the two lurid companies were mixed and released together into a natural pool, they quickly sorted themselves out into schools of like-coloured brothers and sisters – red with red and blue with blue. To show that the tadpoles were not simply exercising a colour preference, Waldman dyed half of one clutch red and half blue – and got schools made up of roughly equal numbers of the two colours.[388]

The tadpoles, it seems, recognise each other by a communal smell, acquired from and learned while still encased in toad egg jelly. This fellowship brings them together later in groups that not only share the same genes, but use a system of chemical communication to regulate each other's growth. When times are hard and food is short, smaller tadpoles secrete a substance so expensive to produce that it kills them, but at the same time gives a rapid and timely boost to the development of their bigger brethren. And as maturing tadpoles turn into terrestrial toads and crawl out of the water, they leave behind them the parting gift of a similar secretion that urges the rest of their gene pool on to warty fulfilment.

This sort of chemical facilitation is by no means peculiar to tadpoles. In 1971, Martha McClintock of the University of Chicago discovered that humans too have such hormonal links. She started with anecdotal evidence to suggest that women who live together tend to menstruate together, and identified certain individual women who seemed to act as "drivers", imposing their cycles on others with whom they shared a room. McClintock simply collected sweat from the armpit of such a leading lady who had a regular 28-day cycle and did not shave or use an underarm deodorant. One drop of her perspiration dissolved in alcohol was placed on the upper lip of a test group of women three times each week for four months. A second control group were treated in the same way with a drop of pure alcohol. When the experiment began, there was a mean difference in menstrual pattern of over nine days. After four months, the control group remained in oestral disarray, but 80 per cent of those who smelled the potent sweat, menstruated on precisely the same day as their otherwise unknown "driver".[260]

We microsmatic animals pay scant attention to such olfactory cues, but they play a very large part in our lives. Tests at the University of Philadelphia have shown that the sex of "odour

119

donors" – who were asked not to clean their teeth, eat spicy foods or wear cosmetics – was easily identified by "odour judges" who simply smelled the exhaled breath at the other end of a glass tube passing through a screen. The average success rate was over 85 per cent, with fewer mistakes being made when donor and judge were of the opposite sex.[83]

In addition to this nasal aid to courtship in the dark, we have a tadpole-like ability to recognise not only genders but genes. At Vanderbilt University in Tennessee, pairs of human siblings were issued with identical white T-shirts which the children wore to bed for three consecutive nights. On the morning of the fourth day, all the shirts were enclosed in plastic buckets with small holes in the lids and each child was asked to identify the one belonging to their brother or sister. Eighty per cent of the children had no hesitation in doing so with a single sniff. And in a follow-up test in which the children's parents put their noses to the problem, 89 per cent were not only able to identify their own offspring, but could also tell the two siblings apart.[282]

An ability to recognise siblings is obviously useful in coming to the aid of close relatives with whom you share most of your genes. It can be equally valuable in helping to avoid mating with them, which could be disastrous. The best interests of reproduction require the choice of a mate similar enough to keep advantages in the family, and yet different enough to avoid the dangers of inbreeding. At Cambridge University, Patrick Bateson has found that Japanese quail *Coturnix coturnix* consistently choose to mate with birds which are similar, but not identical, to the ones they were raised with. The birds recognise nest mates by sight and sound and exercise a delicate aversion to them, which means that under natural conditions quail nearly always end up choosing their cousins instead.[13]

Our incest taboo appears to work in much the same way, making us reluctant to mate with anyone we grew up with, regardless of their genetics. One study of 2769 marriages in Israel found that only six of these took place between people of the same age group in any one kibbutz, despite the fact that such unions were actively encouraged by the authorities. Familiarity, it seems, makes recognition easy, but also breeds a certain amount of selective contempt.[342]

There are clear and constructive genetic influences in action in such sensitivity, making it possible for animals and humans to help their relatives, avoid incest and care for their own offspring. All such responses provide a useful cement for social behaviour, maximising what pioneer sociobiologist William Hamilton called "inclusive fitness" – which is the sum total of the reproductive success of any animal, *including* that of all its relatives.[144] But there is something else, a totally non-genetic influence, that brings individuals together in groups large enough to have powers quite unlike anything available to families or associations held together only by chemical bonds.

Such groups become organisms in their own right, with a separate and intriguing natural history. I am just beginning to appreciate how rich this is and am indebted to philosopher George Maclay for drawing my attention to the antiquity of the idea.[264]

According to Maclay, Plato in the fourth century BC, was the first to think of human society in an organic way. The Socratic dialogue we know now as the *Republic*, was originally entitled *Politeia* – which translates more precisely as "The Being We Call The State" – and is, in essence, an argument for looking at the ideal society as a perfect person, as a giant with a man-like mind.

Plato the idealist was followed by his pupil Aristotle, who was more of a realist – and a great naturalist. No creature was ugly in Aristotle's eyes. He saw beauty everywhere in the animal world and as he came to know its detail better, realised that there were patterns in it too. A tendency for things to come together, to take on forms that had strength and meaning. He suggested that there was a hierarchy of form and that structures such as leaves were subservient to the essential wholeness represented by a tree, which could stand alone. An oak, he said, does not have to be part of a forest to be an oak. A forest is merely a collection of trees. But a man, he decided, is different. "Man is by nature a political animal" and cannot live alone. He needs other people and wholeness in our case is the form of the society upon which we are dependent. In other words, human society is not just a collection of people, but part of nature's plan – an independent form with its own natural history.

This very Greek insight found clerical favour in the beliefs of Saint Augustine and Thomas Aquinas, but it was lost to secular philosophy for two thousand years until it surfaced in Europe during the Age of Reason. Part of the process of new enlightenment was the English philosopher Thomas Hobbes, who in 1651 published an analysis of society which he called *Leviathan* and described as a mechanical giant. Machines were just beginning to transform social life and weigh heavily on people's minds, so it came easily to Hobbes to liken money to a fluid that oiled the joints of his monster, and government to a set of strings and pulleys that moved its limbs into the required positions. His only concession to the Greeks was to suggest that man had constructed this giant in his own image as a self-interested sort of creature; but it was, he insisted, one with an artificial, not a natural integrity. It had power and influence, but no soul.

This bleak, mechanistic philosophy from early industrial England was replaced in the mid-nineteenth century by a more vitalistic French interpretation. Auguste Comte emphasised social bonds and described human society as something very different from the men and women who lived inside it. It was, he thought, a living being governed by its own set of natural laws, and essentially benign. Comte was an unshakeable optimist, replaced towards the turn of the century by a countryman who, although sharing Comte's belief in society as a living organism, suspected that the creature had a more alarming side.

Émile Durkheim was the father of modern sociology. He spelled out in fine detail how our minds are conditioned by the needs of society, which he described as a collection of organisms that appear in a variety of guises, but always treat humans as biological pawns in a grander game. "History," he said, "is *their* story, not ours." They rule us with laws which we accept and pass on to our children, under the illusion that such things are actually under our control. Individual minds can form groups, acknowledged Durkheim, but sometimes in the process they give birth to psychic individuality of an entirely new sort. They produce a being capable of creating ideas or "social facts" that are beyond us. The whole is always more than the sum of its parts. Just as life appears, like magic, when a collection of inanimate molecules are arranged in a certain way, so society transcends the individual. In his *Rules of the Sociological Method*, Durkheim assumes a discontinuity between psychology (the mind of the individual) and sociology (the mind of the society). "The group," he insists, "thinks, feels, and acts quite differently from the way in which its members would or could if they were isolated."

It is to the group mind that we have to look for the source of things such as moral rules, standards of taste, fashion, tradition and popular proverbs. A solitary human has no need of laws prohibiting murder, rape or tax evasion; enjoys no awareness of the rights and duties of parents or partners; feels no shame in being ignorant and cowardly, or any pride in being bright and brave; cares nothing for manners or being properly dressed; and knows nothing about trade or technology, art, literature or language.

In short, human society is a product of evolution. It is created by natural selection and environmental pressures which bring individuals together in a special and powerful way, but it requires no physical change or mutation. It is a composition of ideas and beliefs – a new and essentially psychic phenomenon. A kind of supermind.

And it is glimpses of this, perhaps, that we get whenever we transcend far enough to experience an awareness of a richer reality than our own.

Communication

Landscapes in the drier parts of Africa are decorated with striking sculptures of smooth red earth. These take a variety of forms, from tall compacted columns to fluted, somewhat baroque, little clay cathedrals. Each one is a monument, a frozen product of social behaviour that breaks through the surface like the tip of an iceberg, testament to an extraordinary, but carefully orchestrated, frenzy going on deep underground. These are the cooling towers, the air-conditioning systems, for fungus gardens maintained by colonies of termites. The structures are designed to channel airflow so efficiently through a labyrinth of tunnels that, despite the season, the temperature inside remains within one degree of 30°C and carbon dioxide concentration is fixed at less than 3 per cent.[232]

The architecture of such mounds is elaborate, the result of long periods of adaptive evolution, a complex solution to the needs of an intricate society. And yet the construction of every one is accomplished by hordes of individual termites, each adding their own little piece of mortar in the one way that will give the final structure environmental relevance. If workers are separated from the colony and put into a container with building material, they begin at first to act independently. Each explores on its own, picking up and putting down little parcels of earth, behaving in a seemingly independent fashion until, apparently by chance, two or three such pellets happen to get stuck together. These then act like a magnet to other workers nearby and several combine their efforts, adding more material on top so that gradually a column starts to grow. And when, also it seems by chance, another such column begins to rise nearby, the workers on each tilt their constructions so that the two eventually meet in an arch – the basic unit of termitary construction.

Entomologists today explain these astonishing feats by the use of computer terms such as "dynamic programming". Mound construction, they say, requires no foresight, no grand design. There is no need for termite overseers with plans in hand, because each step of the operation determines the action which follows. Coordination is achieved by perception of what has already been accomplished and the structure grows because small units of fixed behaviour accumulate and acquire wider meaning as a result of a "multiplier effect". In other words, gothic cathedrals take their characteristic soaring forms simply because they happen to be made of bricks of a certain shape or by bricklayers who only know a limited number of ways of putting such pieces together.[417]

Three quarters of a century ago, a more courtly, less mathematical naturalist came to a rather different conclusion. He was an enthusiast with a nimble mind, someone who kept on asking probing questions. He wandered through the African veld, dressed always in immaculate white, touching things with his cane, poking little holes in termite mounds and watching the reactions of the residents.

We see them appear one by one from the dark depths, each carrying a tiny grain of earth. Without the least thought, each worker rolls the pebble round and round in its jaws. It covers it with a sticky mucilage, sets it in position in the breach and vanishes again into the depths. No reasonable person can imagine for one moment that every small worker is conscious of the purpose of its work, that it carries in its mind the plan, or even part of the plan of the building operations . . . Its work is naturally due to instinct, but it is not the instinct of the worker. It is the instinct and design of a separate soul situated outside the individual termite.[236]

Not content with such observations, the naturalist set up his own experiments. He made a large breach in a termite mound and then drove a steel plate through the centre of the wound, dividing it and the entire surface termitary into two separate parts so that the builders on one side of the breach had no direct contact with the work or the workers on the other. "In spite of this the termites built a similar arch or tower on each side of the plate. When eventually you withdraw the plate, the two halves match perfectly after the dividing cut has been repaired." He knew that the termites did not always build the same kind of arch, because a dozen different widths and curves are visible near the surface of any large termitary, and decided that they were not influenced only by the character of the wound or by the work already done. He tried driving in the steel plate first and then making different kinds of breach on either side. Still the termites built identical and matching structures. "We cannot escape the ultimate conclusion," he said, "that somehow there exists a preconceived plan which the termites merely execute." But where, he asked, does this preconception exist?[236]

This charismatic figure who walked alone in Africa and published in a language that few understood, was Eugène Marais. He was the direct spiritual heir to Aristotle and Durkheim, someone capable of understanding both nature and human nature well enough to draw appropriate social conclusions. The dramatist Robert Ardrey described him as "a human community in one man. He was a poet, an advocate, a journalist, a story-teller, a drug-addict, a psychologist, a natural scientist . . . he was unique, supreme in his time, yet a worker in a science yet unborn."[2] The science is now known as ethology and Marais made no direct contribution to it, but his ghost continues to haunt the discipline, dancing on dark nights in a surprising number of academic corridors.

Marais' major concern was the evolution of the human mind. He abandoned his law practice in 1910 and retreated to a remote mountain area of the western Transvaal, where he immersed himself in the wild, watching baboons, birds and termites, and doing his elegant experiments. He began in 1923 to publish a series of speculative articles, written entirely in Afrikaans and appearing only in local newspapers, in which he came to some stunning conclusions. These were new and radical and might well have had an influence in Europe, where Freud had already revealed the personal unconscious and Jung was describing social and collective forces inherent in it. But Marais was half a hemisphere away and half a century too soon, and it is equally likely that he would simply have been ignored.

It is easy now, with full knowledge of all the recent insights provided by comparative ethology and sociobiology, to see how prophetic Marais was. He stressed, for instance, the variety of form in "white ant" society and suggested that each of the peculiar castes was directly analogous to the organs in a vertebrate body. Termite gardeners represent the body's digestive tract, he surmised; the soldiers act as its arms; workers behave like cells in the bloodstream, transporting nutrients and news; and the queen is the reproductive and endocrine system, churning out new tissue and sending out the direct equivalent of sperm and ova on nuptial flights. There are, Marais pointed out, insects very much like winged termites that still lead essentially solitary lives. But somewhere along the line, something new was added to these lives, an amalgamation occurred, specialisation took place and a composite creature with totally new talents came into being. A colony of termites, he decided, is in effect a single organism living in a structure which must be seen, "not as a heap of dead earth, but as the body of a separate and perfect animal".

Marais was well aware that this creature owed a great deal to coordination provided by the queen. He showed how removal of a queen brought all work in even a badly damaged termitary to a halt. And was quick to equate such automatic control with the human unconscious, calling it our "old animal psyche in a state of inhibition" and suggesting that it was part of our natural inheritance which could however produce psychological disorders when released in the wrong way. He acknowledged the reality of instinct, but suspected that it was not the whole story, that there was also something else that operated in all social creatures from ants to people – an intangible sense of identity which he called "the group psyche or soul".[235]

The voice of Eugène Marais was not the only one crying in the wilderness. In the same year that he began his mountain vigil, a remarkable lecture was given at Woods Hole in Massachusetts by William Wheeler, one of the last great general naturalists. He came to the celebrated Marine Biological Laboratory saying that he felt like "some village potter who is bringing to the market of the metropolis a pitiable sample of his craft, a pot of some old-fashioned design, possibly with a concealed crack which may prevent it from ringing true". What he brought, in fact, were some profound new insights.[406]

An organism, said Wheeler, is not a thing, but a process in continual flux. It may be as simple as a bacterium or as complex as the universe, but it has a distinct and changing character involving idiosyncrasies of composition and behaviour. He chose an ant colony as his example, starting off by pointing out that the shape of the nest is as unitary and as adaptive as the skin of any living creature. Many ant mounds orient themselves to the Sun, showing definite tropisms, turning like plants to the light. They begin with a fertilised queen – he called it "a winged and possibly conscious egg" – and grow into colonial organisms which reach maturity only with the production of new sexual forms. Individual workers and soldiers, like cells in a body, seldom survive long, but the life span of the whole ant organism may be forty years or more, during which time it is governed by an agency he called "the spirit of the hive". Wheeler decided that "one of the fundamental tendencies of life is sociogenic. Every organism manifests a strong predilection for seeking out other organisms and cooperating with them to form a more comprehensive and efficient individual." And in eerie anticipation of sociobiology, he concluded that such organisms show a far-sighted ability "to secure survival through a kind of egoistic altruism".

Wheeler, on another occasion, referred to legionary ants as:

the Huns and Tartars of the insect world. Their vast armies of blind but exquisitely cooperating and highly polymorphic workers suggest to the observer who first comes upon these insects in some tropical thicket, the existence of a subtle, relentless and uncanny agency, directing and permeating all their activities.[405]

I am party, seventy-five years later, to research on chemical signals which coordinate the migration of army ants, synchronising their movements with reproductive cycles.[330] I can see how group raiding techniques and nomadic behaviour have evolved to give such new and larger colonies a wider choice of food sources. But I have stood in one of those tropical thickets in Brazil and watched a bivouac of half a million ants clinging together in a musty-smelling mass, slowly dissolve at dawn, flowing out in a fan-shaped swarm that feels its way out across the forest floor, splitting up and recombining, flushing out spiders, beetles, wasps, snakes and fledgling birds, cutting these to pieces and absorbing them all relentlessly into its fabric. And I know precisely how Wheeler felt. There is something uncanny about the cohesion. It is difficult not to see the army as an organism in its own right, as a creature with its own peculiar structure and voracious appetites.

This is a very real dilemma in biology. At what point does such a society become so perfect that it ceases to be a society at all? How do we define an individual animal that cannot live apart from that society? In a very real sense, a honey bee is not an organism at all, but a totally artificial human concept. An individual bee is sterile and no more capable of reproducing itself than an isolated red blood cell. Identity in the bee business belongs to the hive, not to any of its constituents. But if the hive is the organism, then does its existence or its personality depend on the number of components? How many of these can be removed from the hive before it can be said to have died? And how many were necessary to give it an identity, to make it an organism, in the first place?[397]

Part of the answer may be found in the fact that societies seem to have a critical mass. When sufficient individual components are present to lift the community over a certain threshold, it changes character entirely and moves from being a random group into becoming a super-organisation with properties of its own. It becomes a new kind of creature which enjoys a new and awesome cohesion, making decisions and acting on these with a speed and precision that belie its independent origins.

The ability of termites in a colony to work towards the same end on either side of a barrier, no longer looks quite so bizarre if you consider them as parts of the same organism – as right and left hands responsive to orders from the same brain. Marais described one termite queen, 2400 times the size of her workers, who vanished from her rock-hard cell to reappear in another part of the palace, beyond doorways and tunnels far too small for her to squeeze through. But even this strange incident becomes less obviously mysterious if you think of her in the same terms as a group of cells, a sort of benign neoplasm perhaps, which is capable of breaking down and reforming in another part of the body.

I am not, I hope, manufacturing mysteries where none exist. The English naturalist Edmund Selous in 1931 suggested that rapid coordination in flocks of birds could only be explained by what he called "thought transference".[334] But we know now, from single frame analysis of high speed film, that normal reaction times can be accelerated to an apparently supernatural extent by "chorus line" effects, which allow flying birds or dancing humans to anticipate the arrival of manoeuvre waves passing through their groups.[399] It is obviously necessary to eliminate all other possibilities before jumping to unusual conclusions, but what I am suggesting is not that outrageous.

It is simply this: That certain kinds of societies, by their nature or by their size, allow the individuals involved to behave as though they were parts of an organism, rather than an organisation. And that in this condition, they have the capacity to communicate more directly with each other, passing messages, making contact and taking decisions in ways that do not necessarily depend on touch, taste, sight, sound or smell.

There is evidence, provided by Marais and others, to suggest that some social insects ought to be considered in this way. But to me the most impressive proof of group transcendence of individual capacity comes from studies of the behaviour of humans in crowds.

Crowds

The scene: a street in any of our inner cities; slightly seedy, with dark façades, barred windows and just one open doorway, that of the fastfood outlet on the corner.

The time: late evening; a few men stand together near the entrance to the store, others on the street walk quickly towards their individual destinations, studiously avoiding eye-contact, skirting the group on the corner at a careful distance.

This is how it usually is. People in this environment live on the edge of panic, creating spaces around themselves, trying not to touch for fear of being misunderstood, snarling or apologising instantly when by accident they do. It is a society in fragment – but tonight is different.

The situation is familiar. High unemployment, rising costs of

129

living, poor housing, ethnic isolation – all with little hope of any improvement – have contributed to disillusion and brought the community to flashpoint. Nobody knows exactly what finally acts as the trigger, setting off the inevitable explosion – there will be endless discussion about it later – but something happens to push things past a critical threshold.

Nothing has been announced, nothing is expected, but suddenly the street is transformed. Windows are thrown open, people come out of doors and alleys, the group on the corner grows and begins to move. More people come streaming in from all sides as though streets had only one direction. There is a determination in their movement which is quite different from ordinary curiosity. Most of them don't know what has happened and, if questioned, have no answer; they simply hurry to be where most other people already are. There is an awareness of a goal before anyone can find words to express it. Before long, everywhere is black with people who seem to have lost all their fear of being touched. They surrender themselves to the motion, pressing fiercely together, finding a strange release in this removal of inhibition, struggling to be where the density is greatest at the heart of the crowd.

At first, the only noticeable property of this composite creature is its urge to grow. It wants to seize everyone within reach. Anything shaped like a human can join it. It knows no limits and admits of no restrictions. It does not recognise houses, doors or locks, and those who try to shut themselves in or deny its hunger, are immediately suspect. The crowd is open and abandons itself freely to its natural urge for growth. It has no clear feeling or idea of the size it may attain. It does not depend on simply filling a known building or stadium. Its size is not predetermined, it simply wants to grow indefinitely. It loves density and the sense of absolute equality that brings. And at first that is enough, but a young vibrant crowd in this euphoric adolescent state is a sensitive thing. The openness which enables it to grow is, at the same time, its danger. As it reaches saturation point, it begins to feel a foreboding of disintegration, it becomes aware of the possibility of its own demise.

As long as a crowd is growing, it feels secure. But as soon as growth becomes restricted, as soon as it runs out of its natural food, it gets irritable and develops a sense of persecution. It becomes hostile and then it starts to break things. Windows on shops and

130

houses, windscreens on cars, are the first to go because they provide such a satisfying sound. The noise of their destruction is equated with the robust clamour of fresh life, with the cries of something newborn, a happy omen of things yet to come. Doors, gates and fences, anything that represents a boundary, become the next target and are torn down and trampled underfoot. And finally comes fire. Of all methods of destruction, this is the most impressive. It can be seen from far off and attracts ever more people. It destroys irrevocably; nothing is the same after a fire. A crowd setting fire to something feels irresistible. It is.

The quiet evening street is now a district in riot, an environment inhabited by an organism that is out of control, but whose life expectancy is low. Such quick crowds are programmed for rapid self-destruction, usually disintegrating as soon as they lose momentum – unless something happens to give them direction. A crowd which has a Bastille or a barrier to storm, enjoys a new lease of life and goes on until this visible and common goal can be achieved. And in the long-term interests of law and order, it is probably best that such a crowd should be allowed to succeed. Because as soon as it does, it loses its reason for being and in the process gives voice to one last triumphant sound before it just melts away. This unique cry, the call of the organism, expresses its unity more powerfully than almost any other action. It is perhaps the most vivid demonstration of the fact that the community of a human crowd is something qualitatively different from the simple sum of its individual parts. And it lives on in the memory of those who hear it, long after the last fire has been extinguished and the streets have been returned to their original state of uneasy neutrality.

This rich vision of a crowd as a biological entity with its own natural history, is that of Nobel Prize-winning novelist Elias Canetti.[44]

He published his fascinating analysis twenty-five years ago in a book called *Crowds and Power*, describing a whole menagerie of crowd species ranging from wild lynch mobs with a quick murder in mind, to carefully domesticated varieties with such long-term goals as a vision of the Promised Land. Canetti's taxonomy includes "rhythmic crowds", whose natural environment is the carnival; "baiting crowds" that hunt in packs; "flight crowds" of refugees; "prohibition crowds", which collectively refuse to do something

MIND

expected of them as individuals; "feast crowds", that gather in moments of rare abundance; and "invisible crowds", phantom organisms consisting largely of the spirits of the dead. They have very different goals, but all begin with what he calls "crowd crystals" around which the organisms precipitate, and each one has recognised and predictable patterns of behaviour that owe little or nothing to the characters of the individuals taking part.

Some kinds of crowd are important to our mental health. We are endlessly inventive when it comes to creating alliances, arranging marriages and forming highly unlikely groups of people. The Conservative Party, the Boy Scouts and the Daughters of the American Revolution, are all totally artificial networks – but each is maintained with awesome dedication. A few such organisations can, if given strong social support for long enough, even become forces with survival value. But by and large, they are meaningless assemblies of no biological significance.

There are other associations however, true Canetti-type crowds, which are far more potent. These can usually be identified by the fact that they develop and display their own peculiar rhythms, which are quite independent of individual origins. As individuals, we are responsive to natural cycles and our lives rise and fall around daily, monthly and annual rhythms. But as a species, we form crowds which recognise totally alien cycles such as the seven-day week. There is no cosmic cue, no external event which identifies every seventh day as a Sunday, but some of us set it aside as a sacred day to be marked by congregation and expressions of unity. The Christian crowd enjoys its unity in carefully measured doses, creating a special time unit so forcefully and effectively that its members can go through their entire lives convinced that time is naturally divided into seven-day chunks. And the rest of us are obliged to follow suit.

Space is subject to the same kind of subjectivity. The Zuni crowd in North America, for instance, identifies seven different zones of the universe – each with its own characteristic colour, symbol and other attributes based directly on the seven corresponding quarters of their pueblo. They construct space-in-general by extrapolating from the particular divisions they experience within themselves, and this concept holds individual intellect very firmly captive, at the mercy of the dictates of the crowd or society-mind.[264]

There are other, less predictable, consequences.

132

In 1980, at an open-air festival near Kirkby-in-Ashfield in the Midlands, three hundred children collapsed. Without warning, they suddenly began to pant, twitch, shudder, vomit and keel over – until the grassy meadow looked like a battlefield. Ambulances carried the casualties to local hospitals, where the majority recovered rapidly and completely. The remainder were released the following morning after further careful observation, and then the recriminations began. An icecream vendor with insanitary equipment, careless crop-spraying by local farmers, underground gas leakages, contaminated water supplies – all were blamed for the epidemic, but no germ or vector was ever found.[184]

Equally mysterious outbreaks take place from time to time in hospitals, where they usually involve staff rather than patients. In 1975, another three hundred people, most of them women and all of them members of the medical and nursing staff at the Royal Free Hospital in London, experienced twitches, pain, giddiness, nausea, sensory disturbances and loss of muscular and emotional control. A medical inquiry left no virus unturned, but found nothing and returned a meaningless verdict of "myalgic encephalomyelitis". After a similar experience in an American hospital, which came no closer than the British to identifying a pathogen, the outbreak there was labelled "epidemic neuromyasthenia". These fine-sounding diagnoses served to save face and cover administrative confusion, but epidemiologists remained unconvinced, remembering a plague of dizziness, vomiting and swooning which broke out at a data processing plant in Kansas in 1972. For two days women working there dropped like flies, but the trouble stopped suddenly on the third day when a canny management announced that it had located the cause, which they described as an "atmospheric inversion" that had since subsided.

It was decided in the end that a convincing case could be made for only one possible cause for all these outbreaks – mass hysteria.

Hysteria has had a bad press. It smacks of childish tantrums and a loss of control, and tends to be associated with having your face slapped by some more level-minded bystander. Hysterical symptoms are regarded as spurious and illusory, the result it is assumed of a personality defect rather than a physico-chemical cause. In fact, hysteria is very real. It may just mimic the symptoms of other diseases, but it does so in ways which are just as painful, debilitating and alarming – and every bit as venerable. Christianity owes its establishment as a formal religion to an outbreak of mass hysteria amongst its first disciples at Pentecost. Something similar, if less reverent, seems to have assailed a group of unfortunate nuns in the seventeenth-century Ursuline convent of Loudun. And there were regular outbreaks around the tomb of the Jansenist theologian François de Paris after his death in 1727.[183]

Recent research by the United States Department of Health, Education and Welfare suggests that mass hysteria is still very common and may account for hundreds of otherwise unexplained outbreaks of communal disorders in factories and workshops, which are usually attributed to occupational hazards such as leaks of toxic fumes or the notoriously poor quality of canteen food. In these cases, the symptoms tend not to be religious, but to mimic instead those of other common disorders such as food poisoning or gastric flu. And yet the triggers which set off such epidemics have nothing to do with viruses or any of the usual sources of disease. People at work or play, often out of sight of each other and without overt collusion or the chance to compare notes, fall prey to the same set of symptoms at roughly the same time. The possibility of deliberate or even unconscious imitation seems to be ruled out by the fact that the victims at Kirkby-in-Ashfield included babies in prams. It is tempting to see the outbreaks as evidence of suppressed social forces becoming manifest in the common body of a crowd, which reacts as a single organism with its own psychosomatic complaint.[185]

While separate outbreaks of social hysterics tend to have their own characteristic symptoms, there is a common thread. Many also include behaviour that is apparently supernatural. A third-century Greek philosopher described victims who had "not felt the application of fire, had been pierced with spits, cut with knives, and not been sensible to pain."[181] An outbreak in the Haute Savoie in the 1850s was investigated by a French government commission who reported that it found girls somersaulting, hanging upside down and springing from treetop to treetop "just as a squirrel or a monkey might have done".[217] The priest appointed to exorcise the devils of Loudun, claimed that the possessed Prioress read his mind more than two hundred times.[178] And the *convulsionnaires* who became infected on pilgrimages to St. Médard in the eighteenth century, are said to have been able, in the midst of their frenzy, to exercise clairvoyance and to read blindfold.[174]

I am fascinated by the things which sometimes happen to people in crowds. I have tried for many years now, all over India, to catch a performance of the famous "rope trick" – even advertising in the *Bombay Times*, with the offer of a substantial reward for a successful performance. So far in vain. But there is at least one account by two psychologists who, with several hundred other people, saw a fakir throw a coil of rope into the air, watched a small boy climb the rope and disappear. They describe how dismembered parts of the boy came tumbling horribly down to the ground, how the fakir gathered these up in a basket, climbed the rope himself and came back down smiling, with the intact child. Others in the crowd are said to have agreed with most of the details of what happened, but a film record which begins with the rope being thrown into the air, shows nothing but the fakir and his assistant standing motionless beside it throughout the rest of the performance. The rope did not stay in the air and the boy never climbed it. The crowd, it seems, was party to a collective delusion.[386]

It happens often enough. In the sixteenth and seventeenth centuries, panics swept through large parts of Italy and France as people there became convinced that they were victims of a colourless, tasteless and odourless poison called *aqua tofana*. Epidemics of convulsive dancing began in the Netherlands in 1374 and continued through most of Central Europe into the sixteenth century, giving rise amongst other things to the Tarantella. Some of this strange behaviour seems to have been caused by grain infected with the ergot fungus *Claviceps purpurea* – which, as recently as 1951, caused weird visions in a small French village – but bad bread is clearly not the only answer. It had nothing to do with collective visions of saints and virgins at Lourdes in 1858, Pontmain in 1871, Knock in 1879 or Fatima in 1917; or with sacred images that have wept and bled throughout the ages; or statues that are reported to have moved, from the crucifix at Assisi that sent Saint Francis out on his mission in 1208, to the restless virgin of Ballinspittle in 1985. And it is I think reasonable to conclude that such shared experience is not necessarily the result only of religious fervour. On the morning of November 3rd, 1888, thousands of sheep near Reading in England were found scattered over 200 square miles of separately fenced fields, all wide-eyed and panting with the same symptoms of what looked like terror.[243]

I am not suggesting by my earlier use of the word "delusion" that any of these experiences was pathological. I have been involved in enough experiments to test the possibility of extrasensory communication of information, to be convinced that something of the sort exists and can be manifest as a real subjective sensation. An idea, for instance, which arises in the right side of my brain is communicated so swiftly to the left that this reflective organ has no reason to assume that the notion was conceived in any other place. Although I have two relatively independent brains, I am aware of only one identity. And if I should receive a telepathic transmission from some other source, even one outside my body, the chances are that I will enjoy the sensation in the same way, at first hand as a personal experience, and not as any kind of intrusion.

I suspect that mass visions are possible and real in precisely the same way, and that much which we take for granted and readily accept as normal everyday experience, could well turn out to be under social influence and actually rather extraordinary.

*

The time has come, here at the heart of this book, to pause for a moment – to take stock and take a stand.

I have been at pains to point out that there are patterns in the cosmos which rise above mere coincidence and which give everything in it form and relevance.

I have tried to show how living things have drawn on the information inherent in these patterns to arrange themselves in meaningful ways.

I have traced the origin of individual identity to electrical cues, followed the evolution of sensation and suggested that life finds fulfilment in association with others of its kind.

And I have begun to look at ways in which such communality and transcendence might work.

It becomes necessary now to take the idea of collective awareness seriously and to give it a new and untainted name.

As a biologist who has watched social insects and seen societies of simple cells like the slime moulds get together in a common cause, I have no problems with the concept of organisms made up of relatively independent individuals. I see no reason to deny that colonies of bees or ants could be organised and regulated by a force which has nothing to do with the recognised senses. And I am not averse to the suggestion that we may be able to respond to information which reaches our consciousness in ways and at levels of which we are unaware.

I enjoy the poetry of "group souls" and "spirits of the hive", but feel that these lilting descriptions have become uncomfortably loaded with mystic baggage they were never intended to carry.

We need a new word to describe common awareness exercised by a group. A term that can encompass both the influence which guides termites to appropriate architectural conclusions, and the force which holds a human crowd together for better or worse. I want a term which makes no unwarranted assumptions about the nature of the connection or restricts action in any way.

And I suggest a return to two of the oldest and most comprehensive Sanskrit roots: SA – which means "together", and has led through Vedic *sam* and Latin *simul* to "same", "similar" and "simultaneous"; and MAN – which means "to think", and grows through Vedic *manus* and Latin *mens* to "mind", "mental" and "memory".

Combined, these roots produce the noun SAMA and the adjective SAMAN – meaning "something which thinks together or is of like mind".

I am aware of the parallel descent from Sanskrit and Greek of *soma*, meaning "the body" – a word now used in biology to describe the substance of an organism excluding its germ cells. In philosophy, soma also identifies the body as opposed to the soul, but *sama* is intended to be less exclusive.

Sama describes those parts of an individual or society which share information, whether they be in the soma, the germ cells or the mind. Implicit in the concept of sama is an understanding that the whole is qualitatively different from its parts and that it can acquire powers which are more or less independent of its substance – as the mind seems to be of the brain.

Sama is a functional rather than a descriptive term, making it possible now to talk about patterns of sama – i.e. those which exercise a collective influence over growth or behaviour; or saman processes – i.e. those which involve the use of such collective awareness.

*

Control

The ease with which ideas become collective in a crowd is impressive.

In August 1914, barely three weeks after entering World War I, a British army found itself in full and harrowing retreat from overwhelming German forces at Mons in Belgium. The British losses were enormous, but they succeeded in retiring in good order and holding their line – partly it is said, because of a ghostly intercession. On the evening of August 27th, reports began to circulate of phantom archers who stood beside trenches on the front line and held the enemy at bay. Before the night was over, thousands of soldiers, including a Lieutenant-Colonel and his entire brigade, swore that they had both seen and heard knights in the livery of St. George and spectral squadrons of cavalry riding protectively alongside their line of march.[16]

This seems to be an experience similar to the Indian rope trick, in which an idea, drawn from the mind of one person or lifted from the traditions of a nation, becomes broadcast to a number of people who happen to be united by circumstances into a biological crowd. A certain amount of reinforcement and social facilitation appear to take place on such occasions, making the experience more real, more evidential, as the number of those involved increases. There may even be a hierarchy of experience with a crowd, as it grows, crossing successive thresholds and gaining access at each new level to sensation and information unavailable to smaller or less cohesive units. If this is true, it could shed new light on some old dilemmas.

One of the most persistent problems in biology is the question of determination in embryonic growth. If a frog embryo at an early stage is split into two, each half develops into a whole frog. If that part of a later embryo destined to become an eye is similarly divided, the remainder forms a whole eye, not half such an organ. If, however, the whole eye primordium is transplanted to the tail end of another embryo, it does not grow an eye there, but develops instead into a kidney or some other organ more appropriate to that region. Living tissue behaves in a determined and assertive way, restoring the norm wherever possible, counteracting the possibly detrimental effects of environmental accidents. This is not surprising, but it is hard to explain in terms of natural selection. It is not a chance process, but clearly one guided by a predetermined plan.

The fruit-fly *Drosophila melanogaster*, hardy apprentice to countless geneticists, has one mutant form which is blind. More than that, it is totally eyeless – a condition produced by a pair of recessive genes. A pure stock of eyeless flies has lost the usual gene for making eyes, but if this stock is allowed to breed unhindered for several generations, a few flies suddenly appear in the population with completely normal eyes. Not something newly invented as a substitute, but the usual ruby-coloured fruit-fly kind. The orthodox explanation for this startling development is that other parts of the gene complex step in to take over, reshuffling and recombining to deputise for the missing eye-maker. Arthur Koestler, quite rightly, reacted to this suggestion with scorn.

> Reshuffling, as every poker player knows, is a randomizing process. No biologist would be so perverse as to suggest that the new insect-eye evolved by pure chance, thus repeating within a few generations an evolutionary process which took hundreds of millions of years.

Replacement of the missing gene, he decided, must be arranged according to a plan operating on levels higher than that of individual genes.[210]

Biologist Sir Alister Hardy was equally uncomfortable with attempts to explain such control away. "Might it not be possible," he mused, "for there to be in the animal kingdom as a whole . . . a general subconscious sharing of form and behaviour pattern – a sort of psychic blueprint – shared between members of a species?"

He pointed out that sensations of light received by each of our eyes reach different parts of the brain, but merge together in the mind, not in any particular group of cells. Our body, in other words, has something that already exists beyond the single soma, something saman that could be in touch with others elsewhere.[147]

Hardy acknowledged the resemblances between his "psychic plan", the "racial memory" proposed by novelist Samuel Butler, and Jung's "collective unconscious". But he was careful to stress that his organising principle would work alongside natural selection. He envisaged two parallel evolutionary influences – an organic stream drawing on sources of information in the gene pool, and a psychic stream arising from the larger lake of shared experience – both acting, together with the environment, to select those individuals best suited to carry on the species. He was well aware of the strength of external selection, but thought that there had also to be a process of internal selection in which changes were screened at an earlier stage. Mutations in this view may be seen as misprints which are subject to careful proof-reading and correction before being exposed to the harsh and critical realities of publication. Something within the organism acts as an editor, eliminating literals, repairing faulty grammar or lapses in style, and letting through only those neologisms which are genuinely creative and have a chance of survival. It is hard to see how any such editorial control could be exercised without reference to the original manuscript, to the blueprint or plan for the species as a whole. Such access, I suggest, can best be achieved and explained by the existence of sama that has such a broad overview.

There is one biological situation in which something like saman guidance and control is already recognised. It was first described in 1917 by mathematical biologist D'Arcy Wentworth Thompson in his eloquent essay *On Growth and Form*, which has been aptly described as "beyond comparison the finest work of literature in all the annals of science".[375] He argued there that organisms are shaped by a delicate balance of forces and that grand transformations can take place as a result of relatively minor adjustments to this equilibrium. D'Arcy Thompson illustrated his thesis with wonderful lattice diagrams that transform familiar long thin fish into equally well-known short fat ones, by simply changing their spatial coordinates. There is no need to alter the basic design. David Raup, of Chicago's Field Museum of Natural History, has recently used a computer to make the same point about the shape of mollusc shells. He can change the helical coil of a garden snail into a flat and common clam by modifying just two of three simple gradients of growth.[131] The point of these demonstrations is to show that living things are connected in intricate ways and that small changes can resound, echoing through whole organisms or even entire societies. A change in the gene for making retinal cells, for instance, can resonate to lens and cornea, altering each in turn in such harmony that the eye retains its integrity and goes on working even better than before. This interdependence would explain the mysterious evolution of the vertebrate eye, whose extraordinary perfection sent cold shivers down Darwin's spine, and could provide a mechanism for the intervention of the sama in somatic history. It might be argued that the ability of organisms to leap from one optimum to another, from periwinkle to limpet with only a small change in regulation, does away with any need for saman collusion. True, in some cases, but there are others in which reference to a higher authority seems mandatory. Our brains are a case in point.

Animals normally make good use of all the equipment they possess. A few have rigid patterns of behaviour which are not always appropriate. Greylag geese, for example, retrieve eggs which have slipped out of the nest by laboriously rolling these back with the underside of their bills – going through the whole performance even after an egg has slipped away again, leaving the goose rolling fresh air. Ducks and geese in general seem incapable of using the common crow or hawk trick of holding something down with one foot while working on it with the bill, though nothing in their anatomy prevents them from doing so. These shortcomings are the result of fixed action patterns produced in response to relatively stable environments and sources of food, but on the whole, animals exploit their capacities to the full, leaving little scope for further learning. We, on the other hand, seem to have been provided by evolution with an organ we still haven't discovered how to use.

Some time in the early Pleistocene, something happened which began the most explosive growth in evolutionary history. Starting about a million years ago, but picking up speed a little later, our brains blew out like balloons, growing in size and complexity at a rate never matched by any organ in any other animal. The enlarged human cortex now sits astride the rest of the old mammalian brain, bristling with potential we have only just begun to apprehend. For some reason, which remains far from clear, the evolution of our brains wildly overshot our needs, producing a bulge so unbalanced that it has been compared to a tumerous growth, but it is apparently not in itself pathological.

As far as we can tell, the whole thing works amazingly well. The growth of art and literature, science and philosophy, are just the first fruits of an early attempt to gauge its prowess and capacity. All of which is awesome and wonderful, but the fact remains that the human brain is a novelty. There has never been anything quite like it before, at least on this planet, and it cannot be explained in terms of natural selection. Nothing in our environment or evolution demanded such a radical solution, but we seem nevertheless to have been saddled with what Arthur Koestler called "the paradox of an unsolicited gift".[210]

Such blatant excess needs explanation. And as none is available in orthodox natural history, it seems necessary to look in unorthodox directions. Perhaps the stimulus which gave rise to such a luxurious overabundance was not one confined to any mutating individual or population, but came instead from a need to cope and deal with the kind of supersensitivity enjoyed by the biggest crowd in history – by pressure on the species as a whole, all at the same time.

The evidence for direct extrasensory contact between human brains is extensive, even if most of it remains anecdotal. Experimental work under controlled conditions has been going on now for over fifty years. Procedures range from the painstaking efforts of J. B. Rhine at Duke University to make coincidence measurable by restricting response to limited targets such as the five Zener cards; to more refined, and certainly more natural, assessments made on sleeping receivers at the Maimonides "dream laboratory" in Brooklyn.[132] Most such protocols produce statistics that tend to support a telepathic interpretation, but their results come from what psychiatrist Jan Ehrenwald calls "flaw-determined" incidents – those that take place for no very good reason, when there is a temporary or local breakdown of the barrier that normally keeps us from being overwhelmed by stray information. As a biologist, I find these card-guessing tests somewhat meaningless. They seem to me to lack significance and survival value.[88]

I am far happier with the incidents which Ehrenwald classifies as "need-determined" – those which occur when communication is vital and ordinary sensory channels are blocked. It is difficult, perhaps impossible, to reproduce such life-threatening situations in a laboratory, but I find the quantity and variety of spontaneous experience from all over the world very persuasive.

Consider just two examples.

In 1930, an English pilot called Hinchliffe tried to make the first flight across the Atlantic from east to west. It was a distinctive attempt – he had just one eye and his co-pilot was a woman. In mid-Atlantic, two old friends of his were on their way to New York in an ocean liner, unaware either that he was making the attempt or that he planned to take anyone with him. Squadron Leader Rivers Oldmeadow was in bed when Colonel Henderson, in his pyjamas, burst in and said:

> God, Rivers, something ghastly has happened. Hinch has just been in my cabin. Eye-patch and all. It was ghastly. He kept repeating over and over again, "Hendy, what am I going to do? What am I going to do? I've got the woman with me and I'm lost. I'm lost." Then he disappeared in front of my eyes. Just disappeared.[120]

It was during that night that Hinchliffe and his co-pilot died in a crash. This experience is typical of many in which the need is clear. The information is important and meaningful, and seems to be aimed at someone in particular – even if there is little that the recipient can do. The second account is similar and though less dramatic, is even more convincing because it suggests the existence of a biological connection at an unconscious level.

James Wilson was a student at Cambridge at the time and in the best of health. Except that one evening he suddenly felt extremely ill and started to tremble. He tried to ignore the feeling, but it grew so severe that he became convinced that he was dying. He went down to the rooms of a friend, who was alarmed at his appearance and produced a bottle of whisky. After three hours, the feeling passed and Wilson felt well enough to go off to bed. He didn't hear until the next day that his brother had died that evening 75 miles away in Lincolnshire.[294]

Both anecdotes involve the transfer of information between people who enjoy a close relationship, but happen to be separated. Normal channels of communication are blocked, but a message somehow gets through. It is sparked by impending danger and imminent death and it jumps a spatial barrier by descending first to older less conscious levels with wider contacts. It succeeds in much the same way, perhaps, as mystic messages from the right brain get through to the left – by making use of an emotional catapult in the limbic system.

If we describe this mystery simply as "information shared without use of the normal senses", we can avoid the term telepathy, which implies conscious awareness of the receipt of such a signal.

James Wilson didn't know what was happening to him. It frightened him. He was in much the same position as a cat being willed by its hidden owner to take a certain path through a maze, or a dog reacting to another observer's awareness of a concealed object, or a single-celled protozoan moving to an area selected by someone at the other end of a microscope tube. All these situations have been shown to produce results at variance with chance, and they combine to suggest that there is some way in which living things can make contact across barriers impervious to the usual senses. The contact is wide, but it is easiest and most meaningful between individuals of the same species, in which it represents a powerful source of social cohesion.

The pioneer primatologist Wolfgang Kohler once said that "a solitary chimpanzee is no chimpanzee."[215] The same is clearly true of human beings, who go to pieces rather rapidly in solitary confinement. But it seems that, in a very real sense, we are never completely alone. There are strange tendrils that reach out, apparently quite independent of time and space, to touch us all, giving us a common experience of reality.

These are, I believe, the influences of what I have called *sama*. In biological terms it can best be defined as a force that provides us, as individuals and as a species, with the ability to maintain important patterns in the face of change. It is what gives us our identity.

Chapter Six

SEPARATION

Mademoiselle Emilie Sagée was highly regarded as a teacher. She specialised in languages and succeeded, in the complex environment of nineteenth-century Livonia, in reconciling the Russian, German, Polish and Swedish heritage of her pupils by giving them all a solid grounding in international literature. And yet, in 1845 at the age of just thirty-two, she was sacked from her nineteenth teaching post.

The directors of the school for young ladies near Riga were reluctant to let her go, but the parents insisted and began to withdraw their daughters. The problem was that the girls could sometimes see two Emilies, standing side by side at the blackboard or eating the same school dinner. On occasion, the extra Emilie would get bored with imitation and just sit quietly in a corner, but often she wandered further afield. She was frequently seen strolling through the school grounds at the same time as the devoted original, quite unaware of the bilocation, carried on with the lesson in front of her fascinated class.[344]

It is rare for *doppelgangers* to be so persistent or quite so evident, but traditions which describe their existence are widespread. One recent survey of over sixty different cultures, showed that all but three of these accepted the idea that some part of the personality is independent and can travel beyond the bounds of the body. The African Azande, for instance, believe that everyone has two kinds of soul, one of which leaves the body when we are asleep. This *mbisimo*, it is said, travels widely, meets up with others of its kind and has all manner of adventures. It keeps these, however, to itself. We have no memory of them on waking. The Burmese call

145

their travelling companion "butterfly" and believe it to be fragile and easily hurt. The Bacairi people of South America say that we have an *andadura* or shadow, which "takes off its shirt" as we fall asleep. It abandons the body and its experiences form the subject matter of our dreams. Some of these we remember and all of them, suggest the Bacairi, ought to be taken seriously as sources of information about the real world.[340]

I am impressed by the ubiquity of such beliefs. They seem to reflect at least a subliminal awareness of contact taking place beyond the confines of consciousness. This is the area in which I believe sama operates, enjoying the wider view of reality made possible by side-stepping our sensory reducing valves. I suppose that some sort of detachment, if not an actual separation, must be taking place, but see no need to assume that this involves a journey out of the body. It would appear to me to be more a case of opening up a gap between sama and individual soul or psyche – which are not necessarily housed "in" the body in the first place. The assumption that mind is body-bound is unwarranted and has already led to some confusion. It would be more useful to think of mind right from the start as already being "out". Viewed in this way, saman awareness and all supersensory, telepathic and trans-cendent experience, simply become a matter of bringing the mind's independent position and perspective to conscious awareness.

An ability to see for ever depends more on a clear mind than on a clear day.

Multiplicity

When Franz Mesmer ran into political difficulties and was forced to close his clinic and leave Paris, he took to touring the French countryside, teaching "animal magnetism" wherever he could. One of those who learned the technique from him was the Marquis de Puységur, who went on to make one of the most vital discoveries about the human mind.

In 1784, a peasant who worked on the Puységur estate at Buzancy was brought to the Marquis suffering from severe in-flammation of the lungs. The nobleman made mesmeric passes over the body of Victor Race to "balance his magnetic fluid", and the man soon fell asleep. This was a common reaction and didn't worry the Marquis, until he discovered that he could not wake the peasant up again. Each time he tried, Victor entered into a trancelike state in which he seemed not to be conscious, but would respond to any command. Asked to sing, he did so with gusto. Asked to dance, he did that and kept on doing it until told to stop. Puységur was fascinated and called the surprising new state "magnetic sleep". This condition of suggestibility is now familiar to clinicians and widely exploited by stage performers, but it was Puységur two centuries ago who realised that it was more than a theatrical curiosity. He made tests on the entranced Victor which convinced him that the altered state involved a consciousness which was allied to, but distinct from, his normal waking state. And he concluded that every human being is effectively a double system, housing an "I" as well as something else.[339]

During the nineteenth century, the technique known as "artificial somnambulism" was further refined by three Scots. It was described in detail by William Gregory, professor of chemistry at Edinburgh; used as an anaesthetic by James Esdaile, then Surgeon to the government of India; and given the new name of hypnotism by James Braid, who was practising medicine in Manchester. All found that it resulted in a loss of personal consciousness and an insensibility to pain. And they agreed that it produced, along with enhanced suggestibility, a new identity and memory.[376]

Almost a century and a half have passed since those pioneering days and the literature on hypnosis is vast, but even a cursory browse through packed shelves reveals an interesting slant. A large part of recent publishing on the subject is given over to a discussion of methodologies – to debates about how hypnosis should work and to models of how it might. Methodologies in science usually surface when the true nature of the phenomenon being investigated is not immediately obvious. They are ways of helping to decide where to start looking. They are, in short, an admission of ignorance – which is only right and proper, because despite intensive effort, we still don't know exactly what hypnosis is.

The chances are that it has a biological origin. Studies on the Malaysian peacock *Polyplectron malacense*, show something like hypnosis in courtship behaviour. The peacock has his famous fan-shaped tail whose upper surface is a shiny, steely blue, studded with dozens of vivid ocelli. He holds this aloft, framing it with his wings, rattling the quills of the longer feathers to hold the peahen's attention as he pivots to face her, tracking her however she shifts and turns with a gallery of staring eyes. It is the eyes, it seems, that have it. They induce in her a degree of fear that leads to tonic immobility and a crouched attitude, which is precisely the one the male needs for successful copulation.[78]

The importance of both eyes and repetition continue through all such Svengali situations into human susceptibility; but nowhere, it seems, are there any unique physiological or psychological conditions attached to the hypnotic state. There is some evidence of a change in the nature of the body's electrical field, but there appear to be no behaviour patterns that can be said to be found only under hypnosis.[291] And yet there is something about being hypnotised that is unmistakable – and useful. It gets to parts that other states can't reach. When Benjamin Franklin, then Ambassador to France, chaired a royal commission of inquiry into Mesmer's activities in 1784 and brought it to the conclusion that animal magnetism was "a delusion and a fantasy", even he was given pause for thought by one satisfied patient:

> If it is illusion to which I owe the health I believe I enjoy [said the old man], I humbly entreat the experts who see so clearly, not to destroy it; that they may enlighten the universe, that they leave me with my error, and that they permit my simplicity, my frailty and my ignorance to make use of an invisible agent, which does not exist, but which cures me.[277]

The "frail" old man needn't have worried. The same invisible agent is still in business: inhibiting dermatitis that normally results from contact with plants like poison-ivy;[182] and producing such inflammations on the skin of hypnotised subjects who have been touched only with a harmless leaf; conjuring up blisters by nothing more than the suggestion of a burn;[198] curing warts;[199] and increasing bust measurement in women by more than two inches.[413] Ernest Hilgard of the Hypnosis Research Laboratory at Stanford University is convinced that such things are accomplished by a genuinely altered state of awareness, but even this influential figure in the field is reduced to concluding that the only reliable way of determining when someone is actually hypnotised is to ask them.[165] We all seem to know and agree on what an altered state is, but run headlong into irreconcilable differences when we try to define it.

Perhaps then, it is enough to accept that such things do exist, that there is an invisible agent and that while experiencing it we have access, not only to greater control over normally automatic bodily processes, but also to new areas of information – some of which are so radically different that they produce drastic changes in attitude and behaviour and have given rise to suggestion of division in, and multiplication of, human personality.

The first to think in such terms was the French philosopher and psychologist Pierre Janet towards the end of the last century. He studied hysterics and eventually came to the conclusion that their symptoms arose in a sort of second self that seemed to have a continuous existence parallel to and slightly below that of normal waking consciousness. On one occasion while a patient called Lucie was chatting to someone else, Janet was communicating, through automatic writing, with her unconscious:

"Do you hear me?" asked Janet.
"No," came the reply (in writing).
"But you have to hear in order to reply."
"Yes, of course."
"Then how do you do it?"
"I don't know."
"Must there not be someone who hears me?"
"Yes."
"Who is it?"
"Someone other than Lucie."
"Oh indeed. Another person. Should we give this person a name?"
"No."
"Yes. It is more convenient."
"All right, then – Adrienne."
"Well, Adrienne, do you hear me?"
"Yes."[67]

This is a fascinating and very revealing exchange. It records the precise moment of creation of a secondary personality, which takes on its own identity as soon as it is given a name. By insisting on a name, the psychologist assisted at the birth, but the rudiments of new individuality were apparently already there, needing only organisation, just waiting to be identified. And all the evidence suggests that, once such a sub-personality is released, it tends to define itself more and more clearly as a separate individual – even to the extent of denying the existence of the first.

The best-known case of psychic subdivision is probably that of the young woman reported in *The Three Faces of Eve*, who was a quiet housewife and mother until she found herself overwhelmed by the personality of a coarse, seductive woman who loved the night life. When she took her complex confusion to psychiatrists at the University of Georgia Medical School, a third personality appeared. This was a level-headed intermediary who actually assisted in the therapy and helped to bring about a sort of truce, a temporary coalescence of all three "people".[372] "Eve" – now known to be Christine Sizemore – became divorced after publication of the book in 1957, and her shaky reconciliation broke down into even smaller parts. She began to produce new personalities in groups of three at a time, each group operating around a fixed pivot, with the second and third characters obliged to come and go through the first one out. Each sub-personality was so well formed, with such distinctive attributes, that one even suffered from a peculiar allergy. Christine's skin broke out in ugly red blotches in response to a fur coat that she could safely wear without any such symptoms when that hypersensitive character was concealed.[346]

I deliberately use the word "concealed" rather than absent, because there is considerable evidence to suggest that all of Christine's amazing family were internal psychic productions. Each seems to have been the result of splits within her own mind brought about by dissociation as a defence against emotional strain. But the problems she faced with her unruly brood are real ones that present a substantial difficulty for psychiatry. A solution to the enigma of multiple personality, and a clue to its treatment, can be found in the astonishing ability of the human mind to create and flesh out its own mini-dramas. It is hard, however, to stretch this purely psychological explanation far enough to cover cases like that of the "Campus Rapist" of Ohio State University.

Billy Milligan was arrested in Columbus in 1977 and charged with four counts of rape. His testimony was so confused that a psychiatrist was called to assess his fitness to stand trial. She happened to be Cornelia Wilbur, who had already diagnosed and treated a girl called "Sibyl" whose sixteen segments she successfully reconciled into a working whole.[412] Billy turned out to be even more fragmentary. He trotted out a series of twenty-four personalities of both sexes and all ages, each so convincing in its own right that the prosecuting attorney withdrew his case against the hapless young host, who was committed to the Athens Medical Centre.

Billy never suffered from the usual human problem of "mixed emotion". He contrived somehow to parcel out his life so that each personality had a separate function and responsibility – one was the keeper of fear, another of pain, another of anger and so on. The one that got him arrested was a lonely lesbian who had a strong desire for contact with women. This division of labour seems to be typical of those suffering from multiplicity and is particularly well described by one of Billy's community.

> Think of it as if all of us are in a dark room. In the centre of this room is a bright spot of light on the floor. Whoever steps into this light, onto the spot, is out in the real world and holds the consciousness. The rest of us can go about our regular interests, study or talk or play. But whoever is out must be very careful he or she doesn't reveal the existence of the others. It is a family secret.[206]

Up to this point, Billy's case was no more than a complex variation on the theme of partition, but further analysis produced some real difficulties. One of the twenty-four characters identified himself as Arthur, spoke with a British accent, and demonstrated an extensive knowledge of physics, chemistry and medicine. He could also read and write fluent Arabic. Another gave his name as Ragen, spoke with a thick Slavic accent, expressed an interest in weapons and was skilled in karate. He proved that he could read, write and speak in flawless Serbo-Croat. All of which were well beyond the capability and opportunity of young Billy Milligan.

Work on the various personalities of multiples has shown that they not only perform as discrete and cohesive entities in all known psychological tests, but even have distinct brain-wave patterns, which cannot be duplicated by accomplished professional actors charged with the task of creating and rehearsing an imaginary character. And suspicion grows that, in some cases at least, the

151

phenomenon cannot be explained simply in terms of a schism within the biology of a single human being.

Canadian psychotherapist Adam Crabtree uses psychodrama as a therapeutic tool. He encourages his patients to devise and act out scenes which can have an emotional impact and help bring hidden elements to light. He is familiar with our amazing capacity to create and enact personalities, and suggests that such "dramatic personation" is normal and healthy, part of our natural talent for responding inventively to new challenges. But he also recognises situations from his own case records in which people lose themselves so far during such dramatisation that they seem to be taken over – perhaps even possessed.[67]

Possession

Recent Philippine history has been aptly described as "four hundred years in a convent, followed by fifty in Hollywood". But there is obviously more to it than that. Long before either Spanish or American colonisation, the islands of the archipelago were a patchwork of well over a hundred distinct linguistic, cultural and racial groups – many of which still survive.

Cagayan Valley in north-eastern Luzon is home to one such community – an Igorot or mountain people who are marked by Christianity and post-war developments, but nevertheless leave all the most important decisions of their lives to solemn rituals that involve animal sacrifice and lead to consultation with the spirits. Communion is accomplished by *aniteras* or female shamans who are now rare, but carry on like gently beating hearts in dying tribal life. It was to meet one such woman that I made the long journey from Bayombong up into the forests of the Cordillera. I spent several bewitching weeks living in the old lady's compound, watching the daily work of weaving and basket making, taking part in the evening rituals of healing and spirit worship. It was an altogether magical time, but one I remember best for my involvement in what I can only think of as a kind of exorcism.

A child was brought to the *aniteras* suffering from a complaint like none I have ever seen. He was said to be ten years old, and from the right side he looked about that age; but from the left, he had the appearance of an aged and diseased dwarf. From the front, you could see a line running down the centre of his body, as though the Hollywood part of his heritage had spent long hours in make-up that morning, doing their best to make one half of his body look like something designed to be exhumed by Vincent Price.

I can joke about it now, but the effect was truly horrible. The hair on the right side of his head was dark and glossy, while that on the left was dank and lifeless. One eye was clear and bright, the other squint and rheumy. Half his teeth were widely spaced and drawn out into fangs by the retreat of bloody gums, and the skin on that side of his face and down his left arm was covered in running sores. He walked slowly and with obvious pain, hunched with every other step over a left leg shortened several inches by a clawed foot. And when he spoke, which he did rarely, it was out of the twisted left side of his mouth in a snarl and in a language which nobody there understood. Nobody except me. I was astounded to hear, in amongst the deep-throated growl, a few phrases in clear and ringing Zulu – the one African language that I was able to speak when I was his age. The words were odd ones and in-appropriate to that situation, but they left me feeling very vulner-able, as though I had just had my pocket picked.

The *aniteras* decided that the child was possessed by *busao*, an evil spirit – which, in the circumstances, seemed like the only reasonable diagnosis. And for three days she worked her wiles on the child, plying him with herbal potions, saturating him with ceremony and invocation. All to no avail. On the fourth night, however, she was otherwise occupied and the boy/dwarf was sitting on the ground next to a fire encircled by a group of elders, frightening me from time to time with occasional obscene twitches. The people and I were talking in reluctant Tagalog, which is no more their language than it is mine, just passing the time. Nobody was concentrating on the figure at the fire, he was not the subject of conversation and he was looking away from me into the flames. Then slowly, one by one, our gazes focused on him, the talk stopped, the air became almost heavy with condensed attention; and suddenly, as if by prearrangement, the old lady was there with us, standing tall on the edge of the circle. She hurled something into the fire, which flared up in a green blaze and she shouted very loud, very angry, a long quick string of words hurled directly at the afflicted boy.

There was a moment of silence, complete silence, then a terrible scream as the child threw himself down on the ground and began to thrash around violently. Again she shouted, and once more he screamed – a searing combination of pain and anger. It was a duel in sound, a pitched battle that raged and grew into a frenzy, and then stopped as suddenly as it had begun as the child hurled himself face down to the earth and lay still with one arm and shoulder in the glowing coals. For a long, awful moment nobody moved, and then the old woman stepped forward, gently lifted the body up and carried it away to her hut. And it was as though she took with it a great weight from our shoulders – a burden that we were not conscious of carrying, but that had been with us ever since the weird child had arrived.

The next morning, the boy was up early with the rest of the women, helping carry water. He looked straight at me for the first time and his eyes, both eyes, were clear. By that evening he was talking normally, in his own tongue, and walking with only the suggestion of a limp. And by the end of the week, his skin and teeth and hair, his whole appearance, were those of any other healthy, unmarked, active and attractive Filipino child.

I make no apologies for telling this story in such detail and without corroboration. I am not offering it in evidence, but as a starting point for a line of argument. Three things about it are of interest to me. The first is the laterality of the affliction – which, however it was caused, suggests at least a biological vector, involving just half of the brain. The second is the nature of the cure – which was both rapid and dramatic, suggesting the sort of catharsis that has mental rather than physical origins. And the third is the use of an unfamiliar language – in the presence of perhaps the only person out of fifty million in the Philippines who could have understood.

I am not claiming that the child was possessed. I discovered later that his problems had begun three years before when his mother was run over by a truck, killed and hideously disfigured as he was walking down the road with her – holding her right hand. There are, however, strong resemblances between this incident and several other accounts in the literature of what has been identified as demonic possession – most notably the case of fourteen-year-old Karen Kingston, who was cured of a similar affliction in North Carolina in 1974 by a group including three clergymen, a psychologist, a psychiatrist and a general practitioner.[276]

Karen's problem also began with a murder. She watched her

154

mother stab her alcoholic father to death with a butcher's knife and retreated into a state of helpless shock, which turned gradually from withdrawal into ugly and violent deformity. She was sent to a home for retarded children and when orthodox medicine seemed incapable of preventing the transformation of this adolescent girl into a malevolent crone, an evangelical Baptist suggested exorcism. The ceremony began, in the presence of several clinical witnesses and nursing staff, with a minister addressing the child directly as though she were possessed by a devil. "In the name of Jesus, demon come forth! Leave! Leave!" And the girl responded in kind, in a deep voice. "This child is mine! Go away! Go away! Leave us alone!"

The rest of the treatment will be as familiar to those in our culture who know anything of the rites of exorcism, or have even seen the film of *The Exorcist*, as the tactics of the *aniteras* were to her Igorot audience. Events moved from blasphemy and poltergeist phenomena to a violent crisis precipitated by forcing the alleged demon to reveal its name. The moment that the symptoms could be gathered together into a separate identity, which chose the surprisingly undemonic name of "Williams", Karen was free of it. And over the next three days, she regained her health and intelligence and control.

I gloss over these events in a paragraph, as though conversation and conflict with a demon were part of everyday experience. For most of us, of course, they are not. None of those present in North Carolina will ever forget what happened, any more than I can ignore the dramas I experienced in the Philippines. But to me the crucial aspect of both cases, is that events were clearly culturally determined. They followed the scenario appropriate to the circumstances, drawing on beliefs and expectations relevant to those involved. The cures remain mysterious, amenable one day perhaps to the liberal tenets of the fledgling science of psychosomatic medicine, but the process was essentially traditional and social. Which is why I believe it succeeded. I suggest that the clergyman who acted as Karen's exorcist, also played the devil's role – just as I somehow contributed a few words of Zulu to the Philippine performance. Neither of us was conscious of doing so, but I am convinced that at some saman level we were involved. We added social weight to an individual dilemma and helped move it to communal resolution.

Which brings me back to Adam Crabtree, the Canadian psychotherapist who is impressed with the way some of his patients behave during "theradrama" sessions. "I have often seen them," he says, "operating from very little information, 'become' the person that they were portraying to such perfection that they even used that person's characteristic gestures and peculiar phraseology." He goes on to consider the way an actor experiences the enactment – "Often he has the sensation of being taken over and losing himself during the dramatization" – and suggests that this talent is more than just skill or telepathy. It involves the *presence* of the person being portrayed.[67]

I think he may be right.

Let me return, however, for the next step in the argument, to those Zulu phrases. Parapsychology has a name for the ability to use a language of which a person has no ordinary knowledge. It is called xenoglossy or "foreign tongue" and comes in two forms. "Recitative" xenoglossy is the utterance of fragments of a strange language, as one might parrot Latin phrases without having any idea of their syntax or actual meaning. And "responsive" xenoglossy, which is something far more intelligent, involving an ability to converse in the unknown language. The distinction is vital.[124]

It is possible to pick up a Serbo-Croatian dictionary and learn some words by rote. It is possible even to study the section in the front that deals with the rules of pronunciation and grammar, and acquire a rudimentary knowledge of these. But this will not immediately transform you into a fluent speaker of Serbo-Croat, able to talk your way around Yugoslavia. There is a gap between the theory and practice of any foreign language which requires just that – practice. You can read all the books in the world about cycling, but you still have to learn to ride by trial and error, by wrapping your muscles and your mind around the problem until the technique falls into place. Then it is yours for life.

The Filipino child was not speaking Zulu, he was practising recitative xenoglossy. There are many similar examples in the literature on spiritism – of mediums reciting the Lord's Prayer in Greek or throwing in the odd word that turns out on later analysis to be Egyptian or even Hawaiian. Some of these borrowings can be traced to a phenomenon known as cryptomnesia or "hidden memory", in which we dredge up information from unconscious areas without being aware of doing so. Early this century, for example, a spirit who called herself Blanche Poynings created some excitement in mediumistic circles. She appeared through a hypnotised woman and provided impressively detailed information about life during the time of Richard II and Henry IV. The case seemed to be one that provided evidence for survival, until it was discovered that there was a novel called *Countess Maud* which included the character of Blanche and all the historical detail; and the book had been read to the sitter as a child.[81]

Latent memory of this kind clearly surfaces on some occasions, but it is far from being the whole answer. I cannot imagine any set of circumstances which could have brought a ten-year-old boy in the Cagayan Valley into contact with Zulu at any stage of his life. Nor am I disposed to assume that he was possessed by the discarnate spirit of a Zulu witchdoctor. It seems altogether more reasonable to assume that somehow, the mechanism is still far from clear, he was able to recite phrases that were familiar to me, borrowing them from my mind as I might purloin a word here or there from Serbo-Croat.

An outrageous conclusion? Perhaps, but there is some evidence for social leakage of this kind.

Alister Hardy once said: "I have myself become convinced of the reality of telepathy from two experiences I had many years ago, but they are anecdotal and of no scientific value; yet to me they are as important and as real as any observation I have ever made in natural history." Both experiences involved a Mrs. Wedgwood in Lincolnshire, who entertained officers from Hardy's regiment during World War I and claimed to be able to "see" things. On one occasion she suddenly said, "Oh, I can see your brother quite clearly. I can see him sitting at a table drawing what I think must be some engineering plan; on a large sheet of white paper I see him painting what seem to be squares and oblongs of red and blue." What she described was exactly what Hardy himself had been doing all that afternoon, entirely on his own, preparing diagrams to illustrate a talk on military history. A year later, Hardy was working as a camouflage officer and spent hours painting a large sheet of white card with a vivid pink distemper and sitting, watching, waiting for it to dry before going out to dinner with Mrs. Wedgwood. Her first words to him were "Oh, what have you been doing? I see a large pink square on the table in front of you."[147]

These accounts, as Hardy said, are anecdotal. Many people have such experiences. I include his here partly because they are his, and partly because he was not only a great naturalist but an accomplished artist, someone with a good visual memory, on whom shape and colour made a strong impression. Strong enough, perhaps, to overflow.

Some of the best results in controlled tests on the transfer of information by apparently extrasensory means, come from the *Ganzfeld* or "whole field" technique, which tries as far as possible to reduce patterned sensory input. Subjects are put into a relaxed position, with halved ping-pong balls over their eyes, in diffuse light, to a background of unstructured "white noise". This withdrawal of attention from the immediate environment seems to help in picking up weak signals.[324] Someone in a dream state is presumably isolated in much the same way and there is a certain amount of evidence to show that sleepers are subject to telepathic impressions.[380] But if I am right about broadcasts having particular biological relevance in crisis situations, then the leakage between organisms ought to be most marked when the transmission is strong – when the need is greatest. As it was for that Igorot child in the midst of his grotesque dilemma.

The nearest laboratory equivalent to the real life situation is perhaps the technique of "sensory bombardment", in which subjects are given a comprehensive audio-visual experience inside a wrap-around cinema screen equipped with quadrophonic speakers. The effect of this combined assault is so overwhelming that most subjects put up their own private censors and slip into an altered state of consciousness. And in the process, some of them seem to gain access to information not normally available to them. They make *saman* contact.

The English philosopher H. H. Price came, I think, closest to a description of what happens in his analysis of shared awareness as a "field of interaction". *Sama* is not a thing or an entity and contact through it does not have an all or nothing character. It is not a kind of knowing, so much as a mixture of fact, feeling and expectation – a shared experience. "Telepathy," said Price, "is more like infection than knowledge." And once infection has taken place, once information has been transferred, its nature and future is governed by the recipient, who controls the course of the disease.

If the child in the Philippines had been able not only to use a Zulu phrase or two, but to make appropriate reply in Zulu to my questions, he would have been showing responsive xenoglossy. Which would have been even more dramatic, and a sound basis for arguing that actual possession was indeed involved. As it was, he made no response at all, but there are those who have apparently done so. In addition to the unexpected Arabic and Serbo-Croatian talents of Billy Milligan, which remain unresolved, there are several other examples.

In 1974, a thirty-year-old lecturer in public administration at the Nagpur University in India suddenly underwent a profound change in character. One day, Uttara Huddar was a quiet, single woman with modern tastes, whose home language was Marathi. The next, she dressed and acted like a married woman, spent much of her time in religious exercises, called herself Sharada – and spoke fluent Bengali. Ian Stevenson, a psychiatrist at the University of Virginia, examined her in the company of Bengali experts, who not only declared her competent in the language, but doubted that anyone from Nagpur could speak it in the way she did, without any trace of Marathi accent. Uttara's family insist that she never learned, nor had the chance to learn, Bengali. Yet she undoubtedly does speak it – at least during those periods, which may last from one day to several weeks at a time, when she appears to become Sharada and has to give up teaching.[353]

A skill of this order, apparently acquired without the usual training and practice, deserves to be regarded as truly paranormal. I can see no way that it can be explained by telepathy, saman contact, social facilitation or access to a collective unconscious. Knowledge of, and active use of, a language are very different things. One brain might have access to the information in another, but it is very difficult to conceive of any way that a muscular skill could be transmitted. Speaking a strange language requires long and arduous training of the muscles of the tongue and lips and it is hard to see how such control could be achieved – even by a brain transplant. And yet . . . I could be underestimating our ability to adapt. There are at least two other well-documented records of people who appear to have suddenly acquired manual skills that require an equivalent degree of practice and control.

In 1961, following the death of her husband, a British housewife called Rosemary Brown began to play the piano well enough to give accomplished performances of difficult compositions. These were unknown to the world of music, but their style was familiar.

So they should be, says Mrs. Brown, because the composers are Liszt, Chopin, Beethoven and Monteverdi, who "take over my hands like a pair of gloves".[34]

In 1968, following an outbreak of poltergeist activity in his family home, Matthew Manning realised that he was somehow responsible and began to exploit and direct the phenomenon. He produced "automatic writing" in a variety of scripts, including Arabic, and even more impressively – a portfolio of "automatic drawings" in the style of Dürer, Picasso, da Vinci, Beardsley and Klee. Although Manning has no appreciable artistic talent in his own right, the drawings are highly accomplished, done very quickly, always in the distinctive style of a particular well-known artist, but are not always reproductions of any known work. "It's not me," says Manning, "I simply switch on the energy."[234]

The sudden acquisition of linguistic, musical and artistic skills presents a real problem for any explanation of the paranormal. They seem to rule out any possibility of telepathy or extrasensory perception – even of the more naturalistic saman kind I have been suggesting. Which would appear to leave us with just two other possibilities: true possession – the actual invasion of an alien entity; or reincarnation – which in the final analysis amounts to the same thing, except that the entity involved is dead.

I have philosophic and biological problems with both suggestions, and intend to deal with these later, but want first to take a closer look at the actual limits of the senses. They can be surprisingly elastic.

Remote Sensing

Once Mesmerism had been thrice condemned by the French Academy of Sciences, it took a brave man to bring the matter up again. But this is exactly what a young lecturer in philosophy and psychology did in 1878 – at the same time moving the study in a new and interesting direction.

Pierre Janet insisted that there was more to being hypnotised than a simple physiological response and determined to find evidence of higher activity. In one fascinating case, he presented a subject with a number of blank white cards and induced the hallucination of a portrait on one of them. He then shuffled the cards, dealt them out on the table and asked his subject to pick out the one with the "portrait". She was always able to do so without hesitation, no matter how well the blank target card was mixed in with large numbers of others, or how often the imaginary picture was shifted by suggestion to other cards.[191]

I have tried to duplicate this experiment of Janet's several times – and it works. The reason it does so is that attention during suggestion becomes so acute that, while imagining a face on the target, the subject becomes aware of the most minute peculiarities in the card itself. Even if the experimenter does not know where the target lies, subjects are able to pick out the appropriate card by tiny markings on it, or by a microscopic defect at its edge, or even by the pattern of the fibre in the paper itself – and will, under hypnosis and careful questioning, identify the cues they use.

The amazing Eugène Marais was aware of such a talent. He called it "hypnotic hyperaesthesia" and suggested in 1922 that the accumulation of intellect and individual memory took place at the expense of our traditional senses. "Hypnosis proves, however, that this degeneration in man is not organic . . . when mentality becomes dormant under hypnosis, the inhibition is removed."

Marais collected twenty empty snail shells that seemed to him to have the same shape, size and weight. He numbered these on the inside and arranged them in numerical order on a table. He hypnotised an eighteen-year-old girl, brought her into the room carefully blindfolded, and allowed her to demonstrate the acuity of her sense of touch by closing her hand over each shell in succession without lifting or moving it. In her absence, the shells were then arbitrarily shuffled, but "brought back still blindfolded, she replaced the shells in their original order, without very much hesitation, and without a mistake." He tested the effect of hypnosis on the sense of smell by getting twenty people each to handle a different object before placing it on a tray. The tray was then presented to another subject, this time a twenty-one-year-old girl, also hypnotised. "The girl, blindfolded, took out one object after another and by smelling the object and the hands of the different people, handed each object back to the person who had handled it first, without a mistake."

And finally, in an inspired experiment, Marais showed that the same hypersensitivity extended to what he called "the sense of locality". He took subjects out on to the Springbok Flats on the edge of Namaqualand, where it is quite obvious that the world cannot be round. Everything stretches out in all directions, across trackless wastes of identical low shrubs, to a totally featureless and perfectly level horizon. In the middle of this vast monotony, Marais found the nest of a sandgrouse – three well-camouflaged eggs in a small depression on the ground – and allowed his subjects as long as they liked to study the surroundings. Each was then taken in a straight line 200 yards away from the nest and a further 100 yards beyond that at right angles – putting them, courtesy of Pythagoras, just 224 yards from the nest. Not one was able to find it again. At least, not until hypnotised, then the subjects walked directly back to the target. In later tests, a fourteen-year-old boy was able under hypnosis to return to his starting point from a mile away – and despite every attempt to confuse him, did so unhesitatingly and in a perfectly straight line. "Even when a series of circles were described, and numberless zigzags and angled courses, he was never in the least doubt as to the exact direction in which the nest lay."[235]

There is no doubt that we possess an astonishing awareness which can, under certain conditions, stretch the known bounds of our five recognised senses to extraordinary limits. Giving us, as Marais suggests, abilities to rival those of one of his favourite baboons who could "at a distance of six miles, over a landscape flickering with mirage, recognise without fail among a group of people, a human friend (Marais himself) to whom he was greatly attached". This hypersensitivity has to be taken into careful account in all experiments designed to demonstrate the possibility of "extrasensory" perception. Our reach is far wider than we realise. But there *are* limits and whenever these are exceeded, we have to look beyond the normal senses for explanation.

In 1886, Janet was joined by the English psychologist Frederic Myers and the Polish philosopher Julian Ochorowicz in a carefully planned experiment. Between them, they arranged for a fellow hypnotist to try to control a subject known as Léonie from a distance. Half the committee kept watch on the hypnotist, while the others took up observation posts outside Léonie's cottage over half a mile away. At exactly 8.55 p.m. on April 22nd – a time chosen at random by drawing lots in his home – the hypnotist began to exert his influence, summoning Léonie to him. At 9.25 p.m., a time when she would not normally have ventured out, the old woman walked down her garden path with her eyes shut and, successfully avoiding street lamps and traffic, made her way to the hypnotist's home, finally opening her eyes only when she held him by the hand. And in a series of 24 further experiments carried out at equally random times of the day and night, the team recorded another 18 successful results as the hypnotist got the subject to do his bidding from a distance.[258]

It was this early French work that inspired Leonid Vasiliev, Professor of Physiology at the University of Leningrad, to begin tests of his own in 1924. Vasiliev was interested principally in how one individual influences another and, starting with direct contact hypnosis, began to move his collaborators further and further apart. Eventually hypnotist and subject were in different rooms, or in separate buildings, or on one occasion separated by the thousand miles between Leningrad and Sevastopol, but still the silent commands to "sleep" or "wake" were being obeyed instantaneously 90 per cent of the time. He began with the notion that he was dealing with an electromagnetic connection, but soon discovered that no amount of metal shielding seemed to be able to prevent the signals getting through. And it was at about this point, in the mid-1930s, that official support and funding suddenly dried up and Vasiliev found himself accused of "anti-materialistic heresy".[384]

The early Soviet state had understandable ideological difficulties with the notion of non-physical action. Orthodox science everywhere still does, but such experiments on distant influence under hypnosis continue to be carried out, with equally successful results. In the 1950s, a German researcher found that if he put salt or sugar on his tongue, his distant hypnotic subject would experience the taste of these at the precise time:

She heard the ticking of a watch held to his ear; she blinked when his eyes were subjected to a flash of light; she sneezed when an ammonia flask was held under his nose; she experienced a needle prick in his finger in precisely the same corresponding location of her own finger.[384]

In the 1960s, there were similar reports from France and England and a compelling film of remote influence in action in Czechoslovakia. But the whole slant and temper of the research underwent a quantum change with a landmark paper published in *Nature* in 1974.

Physicists Russell Targ and Harold Puthoff joined forces at the Stanford Research Institute in California in 1972. They built themselves a visually opaque, acoustically sealed, electrically shielded, double-walled steel isolation chamber deep in the heart of their laboratory. Into it they placed a man, the "receiver", wired up to an electroencephalograph. In another room they put a second man, the designated "sender", facing a photostimulator – an instrument designed to produce patterns of bright light which flashed into his eyes in trains ten seconds long. He was also attached to an EEG machine and the pattern of his brain-waves soon began to flicker in synchrony with the lights. The "receiver" was never conscious of any change, but soon after the "sender" established the photo rhythm, there was an equivalent shift, a sympathetic response, in his EEG, matching the rhythm of his brain-waves precisely to a visual display that his eyes could not see. The experimenters concluded that "a person can perceive a remote state of affairs, even though his perception might be below the level of awareness."[363]

The importance of this experiment is not only that it succeeds in demonstrating that people can share events to which they have no direct anatomical access; but that a capacity for such perception could be latent in all of us. It gives things like telepathy and extrasensory experience a firm biological base. And it makes it seem somehow less absurd to assume that individuals, even those widely separated in space, can be linked at a saman level and, in effect, share the same sense system.

Targ and Puthoff went on to explore distant connections in a more naturalistic way. They developed a procedure which involved pairs of researchers, one of whom would act as a "beacon" – going out to a target site chosen at random from a list of sixty, and staying there for fifteen minutes, while the "viewer" back at base concentrated on trying to pick up information of the partner's experience. The "viewer" gave a verbal description, and sometimes a sketch, to an interviewer, and an independent judge was then taken to all the listed sites to see if any matched the drawings or descriptions. They did so surprisingly often – on average 66 per cent of the time.[364]

This technique for "remote viewing" has now been replicated in forty-six separate studies at places such as Princeton University, Mundelein College in Chicago and the Institute for Parapsychology in North Carolina. [362] Distances involved varied from half a mile to several thousand, including on one occasion a "viewer" who worked several hundred feet underwater on a submarine,[326] and the overall success rate – that is the number of descriptions by "viewers" which were judged to be appropriate to sites visited at random by their partners – was around 50 per cent.[367]

I find this astonishing. The tests are designed only to deal with pictorial information coming from a person who is anxious perhaps that the experiment should succeed, but under no environmental pressure. There is no biological need to speed a message on its way. And yet, despite the emotional and mental noise which must be present at both ends of this tenuous connection, between people who do not necessarily know or mean anything to each other, information still gets through half the time and in sufficient detail for it to be meaningful to a third party. Something amazing is going on.

Those doing the studies say that the best results seem to come from viewers who are "relaxed, attentive and meditative", excited by the task, which they see as an adventure rather than as a test of their worth. Viewers are told not to try to analyse their impressions, but simply to say what they are feeling or experiencing at the test time. There is no doubt that something is lost in translation of such feelings to words – in transfer from the right to the left brain. Viewer drawings tend to be more accurate than their verbal descriptions. Successful viewing seems to involve an intuitive perception of general form rather than precise detail – viewers often provide astonishingly good diagrams of the shape of buildings, but get the number of columns or windows wrong. And experienced viewers, those who have had enough feedback from successful experiments to see a pattern in them, say that the best impressions are clearly recognisable as such at the time. They have a gentle and fleeting form, starting with vague impressions that gradually evolve into an integrated image which always comes as a surprise, because it is so clear and so clearly elsewhere.[362]

So far, so good. As a biologist, I have no great difficulty with the concept that individuals of the same species could share at least part of their experience, even at a distance. But now we move into a problem area.

In 1978, the Stanford Research Institute team began tests of a different kind with a freelance photographer called Hella Hammid, who had been found to be a particularly successful viewer. A large pool of objects ranging from a steam iron to a banana were handed over to an independent person who, at random, placed one of the objects in each of sixteen numbered boxes in a locked room. At the Institute, an interviewer simply gave Hella the number of one of the boxes, again chosen at random, and asked her to describe what was in it. Her descriptions were given to a neutral judge to compare with the original objects. In six such tests, she scored what were judged to be four direct hits: describing a trumpet as "a gold, bell-shaped object, brass"; and a rag doll as something "velvety, pliable and floppy, with knots of material and midcalf boots with striped socks".

To avoid the possibility that Hella was getting her information, normally or paranormally, from the person who put the objects into the boxes before the test, a second series of experiments was done with smaller objects sealed in metal film containers. These were shuffled, kept in a safe, and selected by a random number generator, so that nobody anywhere knew what was in a specific can. And yet Hella still succeeded in scoring four out of five direct hits: describing a map pin as "silver coloured, thin and long, with a nail head at the end"; and a leather belt as precisely that, "the image I get is of a belt." And to push her talent to the limit, Hella was asked to try and "see" a series of posters reduced to microdots and double-sealed in opaque envelopes, one of which was chosen at random for her to focus on. She again hit four out of six: describing a Swiss alpine scene as "a snow covered mountain encircled with highways"; and a photograph of an open human hand as "a round thing with five points".[361]

This is not telepathy of any kind any more. There seems to be no other mind involved. This is clairvoyance – an ability to acquire information from a distant and inanimate source – and you can see why that could be problematic for a biologist. There are no brains, no sense systems, no theoretical sama to lean on.

Some parapsychologists go even further and suggest that precognition may be involved. That Hella, or anyone showing such talents, could be getting her information from the future, looking at the actual object when the can or envelope is eventually opened to confirm impressions of it. Logically, they are right; but biologically, I find this even more difficult to accept. It may well be true, as Einstein suggested, that "the distinction between past, present and future is only an illusion", but until we can travel through time at will, that explanation remains one more relevant to theoretical physics than it is to practical biology.[92]

It is also unparsimonious, because we already have some evidence to suggest that it is possible for a subject, without physical movement, to succeed in travelling through space.

Beyond the Body

A young man lies on a simple cot in a soundproof room. He closes his eyes and takes several deep breaths, then subsides into a pattern of shallow regular breathing. Instruments attached to his body show a slight increase in pulse beat as his heart accelerates to compensate for reduced blood pressure. The electrical potential of his skin decreases. His brain-waves slow to a steady alpha rhythm and from time to time his eyes flick rapidly from side to side, but there are no signs of sleep as such. He is in a normal relaxed waking state.[155]

Stuart Blue Harary calls this his "cool down" procedure. He regards it as an essential part of preparing to do something which he and a few others can do on demand under such controlled conditions. When Harary feels that everything is right, he says, "Now. I'm going." The twelve channel polygraph registers an increase in both heart and respiration rates, implying a greater degree of arousal, but his skin potential slides even further into a state of deep relaxation. Two minutes later, all these signs reverse themselves and Harary says out loud, "That's it. I'm back." And then he "cools down" and goes through the whole performance again.

There's not a lot of excitement in it for the scientist monitoring Harary's behaviour, but half a mile away on the other side of Duke University in North Carolina, there is considerable interest. A kitten called "Spirit" is confined to a large wooden box whose floor is marked off into twenty-four numbered squares. He is a very active and unhappy kitten, wandering all over the board meowing frequently. But twice during the 40-minute observation period he stops moving and sits quietly in one square for two minutes at a time – precisely those times when Harary at the other end of the campus claims to be "out".

Stuart Blue Harary, who seems now to be called Keith, believes that he can leave his body and travel. He describes it as feeling like floating in space, sometimes in body form, sometimes as a ball of light, occasionally as a pin-point of concentrated awareness. Always he sees and feels and hears things and sometimes others are aware of him being there. In these tests at Duke, his instructions are to go to where his kitten is confined, to comfort it and play with it. A battery of instruments, including thermistors, photo multipliers and devices for measuring electrical conductivity and magnetic permeability, are scattered strategically about the target room. None of them registers anything at any time in the test, but during those four minutes out of forty when Harary claimed to be out of his body, the kitten's behaviour changed dramatically. "Spirit" meowed 37 times during the control period, but not once when it seems that it was getting a ghostly grooming.[253]

The literature on ghosts and apparitions is filled with cats whose hair stands on end and dogs that growl or cower in a corner for no apparent reason. Robert Morris, recently appointed to the Koestler Chair in Parapsychology at Edinburgh University, once took a whole menagerie with him into an allegedly haunted house in Kentucky to test their effectiveness as ghost detectors. He chose a dog, a cat, a rat and a rattlesnake and sent each of these occult equivalents of the miner's canary with their owners into a room in which a murder had occurred. The dog refused to go beyond the threshold. It snarled and "no amount of cajoling could prevent it from struggling to get out". The cat leaped up on to its owner's shoulder and "spent several minutes hissing and spitting at an unoccupied chair in the corner of the room". The rat remained unmoved, but the rattlesnake "immediately assumed an attack posture facing the same chair". None of the animals responded to any other room in the house.[252]

I have dealt with the problem of survival elsewhere, and mention hauntings here only as a reminder of the relative acuteness of the sense systems of other species.[397] It is probably no accident that witches are so often associated with cats and owls – two animals whose choice as "familiars" suggests that they may be very useful for picking up subtle signals. It could be productive to join forces with other species far more often in this kind of investigation, but sometimes it is not necessary to have any intermediary at all.

Robert Monroe lives in the Blue Ridge mountains of Virginia. He used to work in the communications business, but now runs an Institute at which he trains others in a technique he discovered twenty-five years ago, which seems to make it possible to travel out of the body at will. His results have been impressive, but perhaps his most interesting achievement to date was to make a disembodied call on a friend holidaying in a country cottage. He found her talking to a neighbour and, unable to attract her attention and despite his insubstantial condition, decided to give her a playful pinch. She jumped and rubbed herself and looked puzzled. When asked later if she remembered anything untoward happening, she replied by displaying a small bruise on the precise spot.[249]

In 1863 the *City of Limerick* ran into a fierce storm in mid-Atlantic. Aboard her was a man called Wilmot, a manufacturer on his way home to Connecticut, who dreamed one night that his wife appeared in his cabin in her nightdress, kissed him and vanished again. He said nothing about the dream, but the following morning his cabin-mate remarked, "You are a lucky fellow, to have a lady come to visit you like that!" On arrival in Bridgeport, the first thing Mrs. Wilmot asked her husband was whether he had received her visit that night. She explained that there had been news of shipwrecks and that she had been worried and tried to make contact with him. She had visualised herself flying, finding the ship, and going to his stateroom. "There was a man in the upper berth who looked straight at me, and for a moment I was afraid to come in, but at last I came up to you, bent over you, kissed you, pressed you in my arms, and then I went away." She was able to describe the ship and the cabin and his cabin-mate in accurate detail.[348]

A little after 2.00 a.m. on the morning of January 27th, 1957, Martha Johnson saw herself travelling from Plains, Illinois, to visit her home 926 miles away in northern Minnesota. She found her mother in the kitchen and—

after I entered, I leaned up against the dish cupboard with folded arms. I looked at my mother who was bending over something white and doing something with her hands. She did not appear to see me at first, but she finally looked up. I had a sort of pleased feeling, and left.

Martha's mother wrote to her daughter the following day to say:

> It would have been about ten after two, your time. I was pressing
> a blouse here in the kitchen. I looked up and there you were at
> the cupboard just standing smiling at me. I started to speak to
> you and you were gone. I forgot for a moment where I was. I
> think the dogs saw you too. They got so excited.[70]

All these accounts, of course, are anecdotal. Each on its own
does not present the sort of evidence likely to convince a sceptic.
But they form part of a spectrum of spontaneous experience
which is large enough and cohesive enough to warrant serious
attention.

In 1886, Edmund Gurney collected 350 case histories of "spontaneous apparitions", which were published as *Phantasms of the
Living*.[142] In 1951, Sylvan Muldoon and Hereward Carrington
concentrated on experience induced by drugs, illness or accident
and collected together another hundred cases in which separation
of a "spirit" seemed to have occurred.[255] In 1954, Hornell Hart
analysed 288 cases of what he called "ESP projection".[154] Between
1961 and 1978, Robert Crookall published nine books with painstaking records of several hundred further cases of "astral
travelling".[69] In 1968, Celia Green made a press appeal that
resulted in 326 reports of "ecsomatic experiences".[134] In 1975,
John Poynton made a survey of 122 "separative experiences".[283]
In 1978, Dean Sheils looked at comparable beliefs and experiences
in over sixty other cultures.[340] And finally in 1982, Susan Blackmore of the University of Bristol gathered all these surveys and
some of her own research into a detailed analysis of the whole
out-of-the-body experience.[24]

The results are hard to dispute. It seems that about one in every
ten people everywhere claims to have left their body at some time.
Most do so only once in their lives, often at a time of crisis, but
about 40 per cent go on to have further such experiences. Some,
perhaps as many as 18 per cent, learn to separate more or less at
will, particularly if the first experience took place when they were
very young. Women seem to do it slightly more often than men.
Apart from accidents and illness, detachment usually takes place in
states of relaxation on the verge of sleep, though it can occur
during intense activity. There are reports of motorcycle riders and

airline pilots suddenly finding themselves floating high above a speeding bike or perched in terror on the outside of a jet aircraft. Most people have less traumatic experiences, describing floating gently above the ground, from where it is possible to look down at leisure and see details, including very often their own bodies. Many find that their senses in this situation are more than usually acute, but sensations are usually confined to sight and sound. Roughly 20 per cent claim to "travel" away from the vicinity of their bodies, and less than 10 per cent seem to succeed in making their presence felt there. Around 14 per cent come back with information that can be checked and when it is, there are many convincing details, but there are also many mistakes and a blurring and blending of experience of the sort that takes place in dreams or distant memory.

There have been several determined attempts to tame these talents in the laboratory, but apart from Harary's success with his kitten, the results – as far as "travel" is concerned – have been disappointing. Charles Tart of the University of California managed on one occasion to get a young girl to "read" the five-digit display on a random number generator placed out of her sight.[365] And Karlis Osis in New York has had some success with a psychic from Maine, who was asked to take a look at a complex piece of apparatus that only makes sense when examined through a peephole that was beyond his reach.[272] Other studies have been equivocal, which is normal for such research under controlled conditions, which seem to squeeze the substance out of natural phenomena. As a biologist, my own inclination is always to go back to the real-life situation in any search for meaning. And the first question that must be asked about any behaviour, structure or phenomenon is "What is it for?" – "Does it have any survival value?"

I suspect that the laboratory version, the induced out-of-body experience, is a pale and effete imitation of the real thing; and that any conclusions drawn from it are bound to be at best, incomplete, and at worst, misleading. It seems to me that the field experience, as described by those who have it in a crisis or near death situation, can be something of enormous biological significance. As an example, I offer the case of William Travis – fighter pilot, author, and advisor to the Sultan of Zanzibar, who now works in the South Pacific, helping establish local fishing industries and enlivening my mail from time to time with wonderful tales from Tonga and Samoa.

> I was trapped underwater, wearing scuba gear, in the cod-end of
> a trawl net whose performance I had been evaluating from the
> net mouth. The trawl was travelling faster than I could swim or
> claw my way out of it; I had lost my face mask and was immersed
> in a slimy, spiny mass of struggling fish; and I went into a
> panic . . . at once I slipped out of my body and was outside the
> net, rear end up, watching myself lapse into senseless struggling.
> The suspended "I" was not aware of being in air or water, of
> being hot or cold. There was no engine, wind or sea sound; just a
> three-dimensional, vivid, panchromatic view. "I" felt a total
> dispassion, no discomfort, worry or fear, time ran rather slowly
> . . . "I" thought of the heavy knife in "its" belt, and watched as
> the automaton reduced its frantic struggling, reached for the
> knife, unsheathed it and sawed a hole in the net through which it
> was able to get out. "I" saw it head toward the surface . . . The
> next thing I knew, I was spluttering and gasping on the surface,
> watching the trawler recede.[379]

The survival value of an experience that can save life, is beyond
dispute. I have no hesitation whatsoever in suggesting that this
function in itself, even if it is never used during the lifetime of most
individuals, gives such separation sufficient biological significance
for the ability to be retained in the population as a whole.

The next question that has to be asked, is "How did it evolve?
Where does such an ability come from?" Here we are on more
speculative ground, but there may be non-human equivalents. One
that springs to mind is the strange passivity of many prey animals in
the very jaws of a predator. Some tarantulas submit in the most
astonishing way to the attack of tarantula-wasps, apparently reconciled to their fate.[398] And one sees something very similar in cases
where a wildebeeste perhaps has been caught, but not killed, by a
pride of young and inefficient lions – who proceed to gnaw the very
legs off the hapless beast, which just stands there staring blankly
into the distance, until it falls down.

I am not suggesting that wildebeeste have out-of-body experiences – though they may well do so. I am however suggesting that
the trance-like cataleptic state of such a victim is adaptive in that it

172

both precludes pain and prevents flight reactions likely to provoke a *coup-de-grâce* from the predator. The chance of such subterfuge allowing a victim to recover and escape may be small, as far as I know it has never been assessed, but it must be better than no chance at all. Given even such slight survival value, the behaviour is bound to persist and, once part of the repertoire, it becomes available for use in other ways later. Perhaps as a vehicle for the kind of separation that permits an individual at risk to take a disconnected and dispassionate view of a life-threatening situation – and find a way out of it.

Susan Blackmore is a pragmatist and argues, largely on the basis of what has been found in the laboratory, that there is no evidence to prove that anything physical actually leaves the body during the separation experience; and she doubts that anything non-physical would be able to have the sort of experiences described. So she concludes that being out-of-body is a psychological condition, an altered state of consciousness, produced by reduced sensory input and drawing mainly on memory and vivid imagery.[24] She could be right. The laboratory results leave a lot to be desired, but I find her conclusion bleak and unconvincing in the face of such a wealth of spontaneous experience – much of which suggests that, even where life is not being threatened, there is a sharing of information between those involved.

Like her, I have trouble with souls and disembodied spirits and survival of the dead. All these things may exist, but there seem to me to be simpler, more natural, less loaded explanations on hand. I am very impressed by the reality of spontaneous separation experiences – I have had one myself and can attest to a vividness which is quite unlike that of dreams or drug-induced hallucination. I think she is right about the experience being psychological rather than physical, but short-sighted in her limitation of out-of-bodyness to individual experience. I suspect that it has become a social phenomenon and that, while undergoing an apparent separation from physical self, we are actually in the process of socialising, of mixing with other selves at a saman level.

If you can accept that individuality is not tightly circumscribed; if you are prepared to concede that, under certain circumstances, we have direct access to others of our kind; if you will allow that this access, while only partial, can be profound – then I see no problems with telepathy, extrasensory or any other separation experience.

173

I am no nearer than anyone else in being able to provide a mechanism, or objective proof of, sama – but I am convinced that something like it exists. I feel sure that, as we arrive at an understanding of the nature of the bonds which hold cells together, as we come to know the forms which regulate the growth of organisms and give societies their distinctive character, we will be in a better position to describe and explain things such as multiple personality, apparent possession, responsive xenoglossy, clairvoyance and out-of-body experiences. They all revolve, I suggest, around a widespread and natural tendency to seek the high ground, to be where the others are, and to join with them in the transcendence that such combinations make possible.

Horace, in one of his Odes, refers to a scattering of people as "an uninitiated crowd". It is only after we come together that we learn the secret words, that we get to hear the Muse and can sing "songs never heard before".

Part Three is an attempt to put words to this music – to explore the paranormal in human experience.

Part Three

PLANET

"This world, after all our science and sciences, is still a miracle; wonderful, inscrutable, magical and more . . ."

THOMAS CARLYLE in *On Heroes* . . . , 1841

There is something odd about Earth. Quite a lot of things, actually, which together add up to produce a picture of a cosmic misfit – a planet that breaks all the rules.

In theory, and in accordance with the laws of thermodynamics, Earth ought by now to be in a state of barren equilibrium. Five thousand million years of energetic decay should have been long enough to reduce the hot young world to a tired old one on which everything has levelled out and settled down. By rights, all possible chemical reactions should have gone to completion, leaving no untidy peaks of excess energy, nothing which could be further tapped or squeezed, all random and wilful disturbance converted to predictable uniformity.

Logic requires a world covered with a thin layer of very salty ocean, totally untroubled by wave or ripple, under a sky without a trace of oxygen, so heavily loaded with carbon dioxide that surface temperatures stabilise near boiling point. The ingredients of life might linger on in the simmering shallows, but chemical energy will have been totally dissipated and nothing remotely like a fire could ever be lit. The third planet from this unremarkable star, ought by every reasonable law of nature, to be in a sober, steady state, null and void and lifeless.

It isn't, of course. Earth bristles with an excess of energy, hums with joyous disequilibrium, and glories in improbable assemblies of highly unstable molecules. The ocean heaves and turns, mixing its ingredients with an abandon that keeps salt concentrations low. The atmosphere is rich in oxygen and nitrogen, violating all the rules of chemistry by leaving this volatile combination uncombined. Rain falls, land rises, winds blow and life blossoms. Anarchy is added to chaos and the result, against all the odds, is order.

The paradox has been apparent for some time, but it seems to be

177

one of those things that loom so large and are so blatantly obvious, that they are difficult to see. It wasn't until 1969 that it was finally given a name. At a scientific meeting in Princeton that year, James Lovelock pointed out that the unlikely composition of the atmosphere, which has succeeded in keeping surface temperatures constant despite the vagaries of sun and time, makes it look more like an artefact than an accident – something created and maintained by life for its own ends. He suggested that atmosphere and biosphere are interdependent, and combine with the lithosphere to form a single responsive organism, the largest living creature in the solar system. A being he named after the Greek Earth goddess – Gaia.

Lovelock sees Gaia as a biological alternative to what he describes as "the depressing picture of our planet as a demented spaceship, for ever travelling, driverless and purposeless, around an inner circle of the sun".[231] It is an elegant antidote to pessimism. By thinking of our world as something living, as a body of which we form an integral part, we can avoid the destructive view of nature as a primitive force that has to be subdued and conquered. We can learn to live with it rather than against it. The Gaian hypothesis also reveals something of the amazing homeostatic way in which things here really work.

Seen from space, Gaia is a colourful misfit; a wispy, white-capped sapphire; very different from its lifeless neighbours. The colour and the life derive from the same source, from the sea that covers nearly three quarters of the surface, softening its edges and soaking up the sun. This heat sink of three hundred million cubic miles of water is arbitrarily divided into oceanic bays that actually flow together into a single tank that is perfectly stirred. If all Earth history is compressed into one "day", the sea is mixed two thousand times in every "minute" of it, distributing warmth and energy evenly round our water-cooled and air-conditioned planet.

Every eighteen "seconds" on this collapsed time scale, the world's rivers dump enough dissolved salts into the sea to double its concentration, but this nevertheless remains around a resolute and reasonable 3 per cent. It is vital that this should be so, because few living cells can survive a salinity which exceeds, even for just a few seconds, a value of 6 per cent. Half the living matter in the world is still found in the sea, and that fact alone seems to make the chemical regulation not only necessary, but possible.

The three dimensional pasture of the ocean teems with life, much of it microscopic, with opal skeletons that drift down as a kind of marine snow into the deeper trenches, where they build up great beds of chalk and limestone. Softer organic parts accumulate in the same way, turning into buried fossil fuels. And the sheer weight of this deposition is great enough to warp the underlying rocks in ways that channel and release volcanic activity, lifting parts of the sea bed back up to the surface, creating warm shallow waters in which corals thrive, constructing barrier reefs and evaporation lagoons, hurling salts in giddy quantity back on to the land.

The details of such gigantic and delicate control remain mysterious, but even without them it is clear that the flesh of Gaia is alive and well; not just adapting to planetary conditions, but shaping and regulating them in ways that are both comfortable and appropriate to cosmic law and order.

Chapter Seven

OBSERVATION

We don't know who discovered water, but we can be sure that it wasn't a fish.

You have to be outside something, able to experience it from a distance, before it makes sense. It is usually the islander who sees the mainland most clearly. Detachment provides perspective, which in turn permits a certain amount of pattern recognition. You get to see the wood only when it becomes too difficult to distinguish individual trees.

This, in essence, is the secret of science. There may be microscopes involved, bringing us ever closer to the heart of the matter; but even microbiology is objective, adding to knowledge by putting space between an object and its observer. This space may be actual – a gap between whale and whale-watcher wide enough to give them both some sense of security; or conceptual – philosophic detachment sufficient to provide a new and revealing point of view. In either case, the division is real and useful; but it can also be misleading.

The problem lies in the fact that the necessary gap is bridged by perception. We experience and understand the world through signals that are received by the senses and interpreted by the brain – and both stages are subject to distortion. The eyes themselves can become selective, ignoring part of what is there, and the brain sometimes insists on seeing things that don't exist at all. We see what we expect to see. We tend to welcome only proofs of what we already know.

Perception is based, to a very large extent, on conceptual models – which are always inadequate, often incomplete and sometimes

181

profoundly wrong. This complex situation arose because signals from the environment itself can be inadequate. The sort of information we need is not always available. And so, knowledge from the past, mixed up with assumptions about that knowledge, which may be more or less appropriate, are used to augment information provided by the senses. Which means that our perception of any situation depends only partly on sensory signals being received at that time. And it is only a very short step from there, to perception which occurs in the absence of all immediate signals and has to be labelled "extrasensory".

It is hard, perhaps even impossible, to define normal sensory perception. Things change so fast that it is difficult not to sympathise with the woman in an Indian court who was rebuked by a judge who told her, "Your testimony is false: last week you gave a different story." To which she replied, "This week I am a different person: if I had given you the same testimony, then *I* would have been false."[45]

Given such manifest uncertainty, the surprising thing is that it should be possible to formulate any rules at all. Einstein felt that "the most incomprehensible thing about the world, is that it is comprehensible" and went some way towards laying down natural law.[92] It is largely to his formulations that we return when we need to define what is "normal". And it is events that seem to contradict those laws that we refer to as "paranormal". But it is well to bear in mind that neither science nor its laws are fixed and paranormality can never be an absolute concept.

Miracles have been defined as "violations of the laws of nature" and as "suspensions of nature by supernatural action". Both definitions imply that the laws are fixed, that nature is not only knowable, but constant – which seem to me to be unlikely and unreasonable conclusions. I see no need to get caught up in the theological dilemma produced by assuming that the laws are hard and fast and that anything which breaks them must be the result either of divine intervention, and therefore subject only to faith; or fraud, and deserving only of dismissal.

Miracles, it seems to me, are timely reminders that science is far from complete. The only trouble I have with them is that they are such rare events. It would be nice if miracles happened more often and more predictably, giving us the chance to understand them better and to redefine the laws they stretch a little more precisely.

But I have absolutely no doubt that they do occur. Experiments, and observations of them, do from time to time run counter to experience and expectation – and we and science are none the worse for that.

Miracles

In the foothills of the eastern Ghats of Andhra Pradesh, there lives an unlikely guru. He is sixty years old now, still tall and slim and energetic, but there is nothing of the ascetic about Sathya Sai Baba. He wears his hair in a frizzy black Afro-style and dresses in flowing silken robes, often scarlet, trimmed in silver. Large crowds gather each day in the courtyard of his ashram and on most mornings he moves amongst them, working miracles.

Sai Baba's speciality is materialisation or, if you prefer, production tricks. He plucks amulets, rings and jewels out of the air and distributes them amongst those close to him. Some of the faithful receive a trickle of scented oil and others little offerings of fresh flowers. On one of my curious visits, I was astonished to be handed a couple of those marvellous syrupy Indian sweets, just cooked and still almost too hot to touch. And on all his walks, Sai Baba continually scatters sacred ash, pouring it out over heads and images, into cupped hands or proffered containers, or on to trails across the ground. This clinging ash is greyish-white, finely powdered and composed mainly of carbon and calcium; but if it comes from any source within his robes, it leaves no trace on the cloth, and it always comes in quantities sufficient to fill huge drums by the end of his rounds.

Watching the guru in action is a little like being part of a multitude in the presence of a messiah. The multiplication of ash is reminiscent of the five loaves and two fishes whose fragments alone filled twelve baskets. Howard Murphet, in fact, tells of a 1970 party at the Ramachandran home near Poona, where Sai Baba succeeded in feeding a thousand guests with provisions intended for less than one hundred. Such miracles of abundance tend to be interpreted as "spiritual feeding", misperception or myth – but they are not that uncommon.[256]

On the afternoon of January 25th, 1949, the cupboards of post-war Olivenza were almost bare. It was midwinter on the tablelands of Spanish Estremadura, food was scarce and there were just three cups of rice left to feed the entire staff and student body of the town's religious institute. The cook, Leandra, put the pot on to boil nevertheless and went on with her tasks, saddened at the thought of having to turn away the poor with nothing to eat that evening. Like any good cook, she kept one eye on the pot and was amazed to find that the few handfuls of rice soon expanded and threatened to spill over the brim of a ten litre kettle. She ladled the overflow into two other large pots, and when these in turn promised to overwhelm her, called the neighbours in to help. For four hours the avalanche of rice went on until 150 people had eaten their fill and most had taken extra rice home to finish later. The mysterious multiplication ended only when the director of the institute finally suggested that, as all the hungry had been fed, the stove should be turned off.[145]

The Vatican's "Congregation for the Causes of the Saints" examined the evidence later and called twenty-two eye witnesses, most of them women from the neighbourhood who had come in to watch the strange event. None of them of course were trained observers, but all were wise in the cooking of rice and testified that, beyond the original three cups, nothing was ever seen to be added later.

This particular feat earned sanctification for the Blessed Juan Macias of Olivenza. Saint Angiolo Paoli was canonised for similar feasts provided for Carmelite missions in the eighteenth century. Saint Andrew Fournet multiplied heaps of grain for the hungry sisters of La Puye at Pitou in the early nineteenth century; and Saint Germaine Cousin is held to have been responsible for dough that rose indefinitely at a convent in Bourges a generation later. These catholic accounts are merely the theological tip of a long tradition of magic cauldrons and fairy cups that could never be drained. None is proof of a violation of the laws of nature, nor do I offer it as such. None is, in itself any more convincing than the performances of Sai Baba – although he at least was subject in 1977 to close investigation by Karlis Osis of the American Society for Psychical Research and Erlendur Haraldsson of the University of Iceland.[146]

I touch on such cases for two reasons. The first is to show that reports of "miracles" are by no means confined to biblical times. Water is still being turned to wine, but these days it seldom gets quite the same sort of press. In 1975, a housewife in Hertfordshire said: "My husband is fond of curries and makes them regularly. About five years ago he bought a tin of ground coriander and each time he makes a curry he puts one heaped teaspoon of coriander in it. Yet the next time he gets the tin out, it is packed tight under the lid. In other words, although he has been using the coriander for five years, the tin is still full! We can't understand it."[243] Her plaintive report earned just two inches in the letter columns of a local paper, which seems unjust; though in all fairness I have to say that it is difficult to know what else to do with such anecdotal experiences. Perhaps all we can do is to keep them and allow them to accumulate in the hope that they will eventually produce some kind of pattern.

My second reason for discussing the subject at all is that it gives me the chance to look at a possibility raised by fellow biologist John Randall. Perhaps, he suggests, miracles are actually quite common after all, and simply go unnoticed most of the time. The feeding of very many on very little is hard to ignore, particularly in Christian circles where it so nicely conforms to biblical tradition. But other less dramatic manifestations can easily be overlooked or explained away as coincidence, illusion or the exercise of unusual skill.[294]

The Duke University collection of coincidences includes the story of a boy and a toy gun. It was one of those antisocial devices that fire nothing, but make a harsh mechanical sound when the trigger is pulled. When it failed to do so one day, the tearful child brought it to his father, the owner of a small country store, to mend. There were several customers in the shop at the time, who joked about it, doubting that he could fix the dreadful gun or that he would seriously want to try. He tinkered with it for a few moments, announced that the job was done and, when his announcement was greeted with jeers, added "If you don't believe it, I'll shoot that clock right off the wall." He pointed the gun, pulled the trigger and the clock leaped from the wall and smashed. There was no wind, no movement, no traffic vibration; all the witnesses later agreed that "everything was still, including us – for a long time."[302]

They probably talked about it for several days and continued to tease the man, but the chances are that sooner or later they will have dismissed it as just one of those strange coincidences. One of those things that shouldn't happen, but goes on doing so anyway.

185

Religious sanction can help to channel and legitimise such coincidence. Each year on September 19th, the feast day of Saint Januarius, huge crowds gather in the Duomo, the principal cathedral of Naples, to witness the display of two tiny ampules in a clear crystal container. These are reputed to contain the dry, clotted blood of the Saint, martyred at the hands of the Romans during the reign of Diocletian. The Vatican has never been able to document his life, let alone his death, and decided to strike the Saint off the official list during the Catholic "cultural revolution" of the 1960s; but this hasn't stopped Neapolitans getting together each year in the hope of yet another miracle.

The sealed casket is suspended out of reach. There is no direct contact with it and no evident rise in temperature, but after some minutes or hours of fervent prayer each year, the dark clotted mass in the phials liquefies, resuming the bright red colour of fresh blood and seeming, on occasion, even to froth and boil. I have watched it do so, and seen the effect on the congregation, who have a strong vested interest in the miracle taking place. "San Gennaro", whatever Rome may think, is their Saint and they know that his failure to liquefy can have disastrous consequences. It has in the past heralded the eruption of Vesuvius, the invasion of Napoleon, and the arrival of the Plague. In 1978 it even coincided with the election of the Communist Party to city government.[145]

I fully appreciate that the liquefaction, in itself, offers nothing to science. It is a small trick, the sort of thing included in most do-it-yourself conjuring kits. History records that it was duplicated even in the seventeenth century and performed frequently for house guests by Raimondo of San Severo, a Neapolitan alchemist and professional doubter – the "Amazing Randi" of his day. So what? The possibility that the phenomenon could be cheated, admittedly exists, but I am more interested in the alternative possibility that it might not need to be. I was in the Duomo during the festival in 1976 and saw it happen. The liquefaction was a little late that year, there had already been rumblings on Mount Vesuvius, the congregation in the cathedral was thicker than ever, and the atmosphere was so electric with communal need that it wouldn't have surprised me to see the crucifix itself dissolve into a puddle.

I am suggesting that this miracle, along perhaps with many others, needs no divine intervention. That we are capable, without outside help, of knocking clocks unconsciously off the wall or exercising enough conscious social energy to change the chemical state of a few grams of some substance in a glass tube.

Religious authorities have long struggled with the problem of miracles, acknowledging the difficulty facing scientists called upon to believe in immutable natural law all week, and to suspend such belief on Sundays. Some manage to do this quite well, exercising a common human ability to believe more than one thing at once. A few great minds have had to wrestle rather longer with the dilemma, coming like Saint Augustine and Teilhard de Chardin to a sort of philosophical reconciliation that sees miracles as events above nature, things of another order than nature, and not necessarily in contrast with it. But this is patently unsatisfactory, providing as James Hansen puts it, "a makeshift designed to let some science get done while someone figures out a better way to put it".[145] Perhaps, in the end, all the mental gymnastics are unnecessary anyway, and have taken place only because we have been slow to recognise the possibility that miracles are home-grown and represent an inherent, if peripheral, part of human biology and behaviour.

Returning to a theme which is, I trust, becoming a little familiar – I point out again that we seldom perceive the world as it actually is. Our brains and our sense organs were developed in the interests of biological survival. They were not designed to satisfy a philosophical need to understand the nature of reality. At each stage in our evolution, we have scanned the broad spectrum of environmental noise and selected from this barrage just those bits necessary to ensure our own survival. We found X-rays and ultraviolet light irrelevant to our immediate needs and ignored them – and they continue to pass by, undetected. Neutrinos are everywhere, vital to the conservation of mass, energy and momentum; but they mean nothing to us personally, having no direct bearing on the more pressing problems of self-preservation and reproduction. Perhaps then, we have a similar blind spot for the mechanics of miracles, which surface into awareness only when given theological impetus or brought directly to attention by debates about someone like Uri Geller.

Biologist John Randall points out that "we all have an internal model of the world which has developed as a result of years of practical experience, and which is perfectly adequate for everyday

187

life." According to this model, we live in a uniform space of three dimensions, in which all objects behave in an orderly way, retain their shape and position unless acted on directly by some outside force, and are lawfully organised in a linear sequence which moves inevitably from the past, through the present and into the future. Our minds are strongly conditioned by this model, programmed and predisposed to look for explanations of each new event which will be consistent with it. And the chances are that any event which doesn't fit and can't be made to conform will be discounted, forgotten or rejected altogether.[294]

Everard Feilding, one of the most acute critics of psychic research in the early part of this century, was aware of the difficulty. He said, "The ordinary effect of the sudden confrontation of a fairly balanced mind with a merely bizarre fact is a reaction: the mind rejects it, refuses to consider it. And the more bizarre the fact, the stronger the reaction." He pointed out that this made scientists particularly vulnerable, more prone than others to react emotionally to events which seemed to threaten their more substantial models of reality; and, as a consequence, more likely either to reject contradictory evidence out of hand, or to be deceived by it.[109]

I know the problem well, having found myself on many occasions torn between intellectual and emotional certainty. Knowing at one level that something was impossible, that it could never happen, and yet convinced at another that it had. Finding resolution of the conflict only in a growing suspicion that the model must be incomplete. Its failings, I suspect, lie principally in its materialistic eighteenth-century origins. It is too rational, too orderly, too sure of itself to be totally true. Life just isn't that tidy. I suggest that there are as yet undiscovered ways in which the inner self and the outer world entwine; that our emotions and beliefs, our conscious and unconscious needs, sometimes find reflection in the behaviour of things in the physical world. And that this happens more often than we realise or can comfortably admit.

Think of this possibility next time you watch a golfer and his gallery will a ball into a hole; or see a tenpin bowler or basketball player or dart thrower track the flight of his projectile, following it with his eyes and body movements as though he were actually in direct and active control of a guided missile. And reflect on the training which makes Zen relevant to the art of archery, produces sharpshooters who are taught never to aim at the target, and turns out champions who first practise their service aces in games of "inner tennis".

Poltergeists

"Timor" in Indonesian means east, and Timor Timor is the eastern end of the largest island of the volcanic Lesser Sundas, which lie like stepping stones between Java and New Guinea. It is a rugged, mountainous area, now in the grip of a lingering guerilla war, but in June of 1974, the capital Dili was quiet, an old colonial town in stately decay.

I was passing through on my way up to a Tetum village to visit a famous *matan do'ok*, literally a "far seer", who was said to be able to make rain on demand. It seemed a good time to test his powers, as the long dry season was at its height, rivers and wells were drying up and the soil was parched and brown. But I never did get to meet the old man, because I was diverted in Dili by news of an unusually active *buan*, an evil spirit, on the coast nearby.

I travelled west from Dili to a savannah shore that was almost African, studded with flat-topped acacias and thatched huts. People there live by making salt, running sea water at high tide into little mud lagoons, letting the sun do most of the work, finishing off the process of evaporation with charcoal fires under huge metal pans. And it was in the home of one of these families that all hell had broken loose.

The household was small, a man and his wife, their two small boys and the husband's young unmarried half-sister – and it was clear to me from the moment I arrived where the source of trouble lay. The mother and father were typical Tetum, dark-skinned people with curly hair, more Papuan in their appearance than Malay. The children resembled them, but the girl was very different, lighter-skinned with almost Chinese features, which in that community would have made it difficult for her to find a husband. They called her Alin – "little sister" – and treated her with courtesy, but it was plain that she was the odd one out and ended up with most of the dirty work.

I ate with the family that evening, or started to, before things began to happen. We sat on the ground around a rough wooden table in the centre of a large grass-roofed room. There was a fire in

189

an open hearth on one wall and a kerosene lantern on the low table. A dish of grilled fish and maize meal had been placed in front of the father and each of us had a clay cup of sweet tea. Alin was busy at the fire opposite when the eldest boy, about eight years old, screamed and dropped his cup, which broke in two on the table beside me. The back of his right hand began to bleed from fresh punctures that suddenly appeared there in a semi-circle, like the mark of a human bite, but with a diameter larger than his own.

As I was examining the wounds, the lamp flame turned blue and flared up, and in the suddenly brighter light I watched a cascade of salt pour down over the food until the entire dish was covered in the coarse grains and completely inedible. It wasn't a sudden deluge, but a slow and deliberate action which lasted long enough for me to look up and see that it seemed to begin in mid air, just above eye level, perhaps four feet over the table. I stood up immediately and walked round to stand near the fire, from which I could see the whole room. Nobody else moved. But the table did.

There was a slow cracking sound, as though the thick wood was tearing itself apart; silence, then a series of sharp raps, the sound of urgent knocking at a door; then it began to wobble. The family got up in a hurry and we all watched in horror as the heavy table heaved and bucked like the lid on a box containing some wild animal, and finally flipped over on its side and burst into flame as the lamp on it shattered on the floor. And, I am ashamed to say, I joined the others and ran.

We stood outside the door for a while and when nothing further seemed to be happening, the father and I went back in and put out the fire. I had by now recovered some presence of mind and enough scientific curiosity to look very carefully at the table, walls and ceiling for any signs of strings, pulleys and other devices. I found none, nor did I really expect to. Everything I had seen could have been arranged for my benefit, but I was certain by then that I was watching a typical, if more than usually violent, poltergeist in action.

My suspicions began in Dili when I was told that the *buan* was a bad one, perhaps even a *swangi* or sorcerer, who smelled of sandalwood, spoke in strange voices and smashed every mirror in sight. I had been involved before with investigations into such "hauntings" in South Africa, Brazil and the United States, and was interested to see if one in Indonesia would have oriental overtones.

It turned out to be very much in the classic mould, part of a global archetype.

I stayed with the salt-makers for the next two days and learned that the troubles had begun during the previous rainy season, when both parents were out working in a newly cleared *ladang*, struggling to get in a maize and tobacco crop, while Alin looked after the children. The boys gave her a rough time, which ended only when a basin of water rose, apparently of its own accord, travelled across the room as if carried by an invisible hand, and emptied over the younger, six-year-old, child. This was followed over the next days by stones that came crashing through the thatched roof, furniture that moved, money that went missing, weavings that ripped themselves off the mother's loom and occasional small outbreaks of fire.

The boys at first had been excited, but I was told that as the activities continued, they began to be infected by the communities' obvious alarm, and grew afraid and withdrawn. By the time I arrived, four months later, everyone in the area had seen some part of the goings-on and nobody would venture near this home after dark. A Catholic missionary and a local healer had both conducted ceremonies around the house, but these seemed to have little effect other than throwing Alin into what sounded like a fit. She had another on my second evening there, showing all the characteristic symptoms of *petit mal* epilepsy, losing consciousness briefly as stones, some of them quite large, began to fly around inside the house in all directions. One, about the size of a walnut, struck me on the chest and when I picked it up, it was hot to the touch. I took it with me as a memento when I left the following day.

I knew by then that there was little I could do to help. These things usually run their course, which seldom lasts more than a few months, and then stop as suddenly as they began. Many seem to revolve around a particular person rather than a place, and are nearly always associated with a tense social situation. I have never heard of an outbreak in a perfectly happy home.

"Poltergeist" is an old German world, meaning "noisy spirit", the sort of thing that went bump during medieval nights, but it has never been confined to Germany. Such disturbances seem to have happened during the times of the Emperor Vespasian and Saint Caesarius of Arles; they were recorded in tenth-century China and twelfth-century Wales; and they appear to be perfectly ecumenical, making life equally difficult for Martin Luther and the Bishop of Zanzibar.[219]

Poltergeist phenomena are by no means uncommon. Police and newspaper files overflow with reports of their activity – there is a useful survey of hundreds collected by the French police between 1925 and 1950.[377] New ones pop up every week in almost every major city. I get dozens of letters each year asking me to come and look at outbreaks in São Paulo, San Juan or Saint Louis. I went the first few times, if I happened to be nearby, but I don't bother much any more. Not because I wasn't impressed – I usually am – but because there is a sameness about them, wherever they happen to be. And that in itself, of course, is interesting.

Alan Gauld, a psychologist at the University of Nottingham, has made a fascinating survey and computer analysis of 500 cases – which he says were not difficult to find. "The chief problem was to stop the accumulation becoming unmanageably larger than that." He classified the cases according to the occurrence of sixty-three listed characteristics and showed, amongst other things, that: only 24 per cent lasted longer than a year; 58 per cent were most active at night; 48 per cent involved rapping sounds; objects moved in 64 per cent of all cases, 36 per cent of these including large items of furniture; doors and windows opened and shut unaided 12 per cent of the time; where an agent or focus was apparent, this was most often female and under twenty years old; just 8 per cent involved detected fraud; a natural cause was discovered in only 1 per cent; and an interesting 16 per cent demonstrated active communication or provided direct responses to those involved.[125]

The last characteristic is an important one. There are persistent and understandable attempts to explain poltergeist phenomena away as earthquakes, ground subsidence, meteorological or electrical disturbances, and the activity of tidal or underground water tables. These natural causes could account for the movement of some items in some cases, but experiments conducted in Cambridge in 1961 showed that this was unlikely. A demolition crane with a steel ball and a special "house-shaking machine" were turned loose on a terrace of condemned houses. It was found that vibration sufficient to crack walls and bring the plaster down from ceilings still failed to move a marble on the floor of one home, or to dislodge a matchstick resting on the edge of a mantelpiece in another. There is clearly more than vibration or any other natural disturbance of ground or air involved in the kinds of movement seen in most poltergeist situations.

Gauld found that major breakages and injury occurred in around 10 per cent of his cases, all without evidence of structural damage to the houses concerned. Animals were disturbed in 6 per cent and another 16 per cent involved some kind of assault – invisible pinches, blows, scratches and bites – on human beings. He could find no evidence that any of the breakages or injuries were the result of fraud or accident, and admitted to being impressed by cases such as a Canadian one of 1889 in which something near Quebec City "took money, spread excrement around, stole food, broke windows, caused fires, cut off hair, threw things, appeared in grotesque forms, developed a voice, swore, blasphemed, repented, became pious, blamed a witch, sang hymns, assumed the figure of an angel and before leaving said farewell to three children." The implication is that, at some level and in some way, consciously or unconsciously, either as a stimulus or as a response, intelligence is involved.

In 1948, Nandor Fodor – an Hungarian American psychoanalyst – noticed that several of his patients with unresolved emotional tensions were associated with, or lived in, houses where poltergeist activity had been reported. He saw some of the phenomena, was satisfied that they were genuine, and took to analysing the poltergeist itself as though it were a disturbed person.[113] His diagnosis was one of fear or guilt that had been concealed, and he suggested that deeply repressed drives or conflicts in individuals with hysteric tendencies, could produce "conversion" symptoms and be externalised in some way – perhaps as classic poltergeist phenomena. This diagnosis was particularly convincing in the case of an adolescent boy who was a potentially creative writer oppressed by lack of recognition even from his family, who soon became the victims of inexplicable disturbances and breakages in their home.[114]

Jan Ehrenwald, psychiatrist at the Roosevelt Hospital in New York, describes the disturbed or rebellious behaviour of children in unhappy families as "loudspeaker" actions, and points out that poltergeist activity in many cases looks very similar.[86] He suggests that the emergence of typical poltergeist phenomena could have a certain restitutive value – at least for the individual who may be "acting out" through them, and who can sometimes be spotted as the one person in a situation who is repressed, paranoid or simply exercises a strong grudge. Ehrenwald concludes that poltergeist, and perhaps other paranormal events, should be seen as "unconscious strategies aiming at alleviating the human situation, if not at self-healing". He wonders, however, about the fact that "the same psychodynamic conflict and family constellation occurs in the best of families time and again," while poltergeist cases are still comparatively rare.[87]

I wonder just how rare they are. The dramatic demonstrations certainly do not occur in every home in conflict; but what family has not at some time found itself nonplussed, and sometimes even at odds, over things which go missing, or are not where they should be, or suddenly seem to be in plain sight when one could have sworn that they were not there a moment ago?

My own experience suggests that the psychological interpretation of poltergeist phenomena is one on the right lines. Every case I have observed has included an individual who fits the Fodor diagnosis or a situation likely to produce such an "agent" – someone unconsciously involved in producing the activities. In the Indonesian experience, Alin was a classic subject in a situation she found repressive, but could do nothing overt about. The limitations of her culture made it impossible for her to rebel or break away, and so she found, perhaps, some sort of release in covert aggression that at least had the effect of inhibiting her boisterous nephews.

Alan Gauld's computer analysis shows that there has been a noticeable increase during this century of poltergeist cases which centre around male rather than female agents, and he suggests this may be because, "these days there are hardly any frustrated maidservants, indeed hardly any maidservants, frustrated or otherwise." He also records a decline in cases in which agents have become the victims of possession, or which have been attributed to witchcraft, or which involve apparitions – and suggests that "It seems likely that these changes are all related, and reflect an influence upon the form of the phenomena of changes in the folk-beliefs in which potential poltergeist agents are likely to be brought up."[125]

It will be interesting to see how exposure to the 1982 Spielberg/ Hooper film *Poltergeist*, with all its angry Indian spirits, affects spontaneous activity in suburban America in the near future.

Psychokinesis

The nice thing about poltergeists, is that they are so accessible.

Anyone prepared to take the time and trouble to track down reports, can sooner or later see or hear one for themselves. They suffer from all the shortcomings of any investigation outside the laboratory – it is hard to establish adequate controls or to be absolutely certain that no fraud could possibly be involved; but they have the merit of being largely spontaneous, of taking place in situations that reflect real biological need, and they often display a degree of very human whimsy.

In 1958, the home of James Herrmann on Long Island, New York was disturbed by a poltergeist that seems to have had a thing about bottles. No cork, cap or stopper was safe. Wine, water, medicine, bleach and shampoo bottles anywhere in the house were systematically opened, spilled and emptied. Gaither Pratt from Duke University went to investigate and discovered that nobody need be near, corks could be heard popping in locked and empty rooms, but that no unusual events took place unless the Herrmanns' twelve-year-old son was home. The boy was apparently not involved in any kind of trickery, but the recognition that he could somehow be the focus, shifted responsibility in the case from religion to science; and when the family had the courage to look at it in this way, the bottle-popping soon stopped.[285]

This seems to have been the first poltergeist case to come under experimental scrutiny by scientists prepared to take the phenomena seriously as physical events capable of measurement, but it was not the last. It led directly to two justifiably well-known and oft-quoted cases. The evidence for both is so strong that I make no apology for outlining each again here.

In 1967 electrical mayhem broke out in a lawyer's office in the Bavarian town of Rosenheim. Lights went on and off, bulbs became unscrewed, fuses blew, photocopiers spewed out their fluids, telephones rang for no reason and made calls that were never dialled. Hans Bender of the University of Freiburg came in to investigate and soon noticed that the events seemed to follow a new employee, a nineteen-year-old girl called Annemarie, beginning as soon as she arrived for work each morning. Lights hanging from the ceiling swung behind her as she walked down the corridors and exploding bulbs showered her with glass. Bender

195

discovered that she had a feeling of resentment towards her employer and he not only decided that she was unconsciously working off this grudge in an electrical frenzy, but determined to find out how.[19]

He called in two physicists from the Max Planck Institute who at first suspected voltage surges in the power supply, but the phenomena continued even after they had isolated the office from the mains and put it on to a controlled and independent generator. They did a series of experiments which, in turn, ruled out the possibility of magnetic fields, ultrasonics, infrasonics, electrostatics, strong vibrations or the presence of any fraudulent mechanism on the premises. All the observers saw drawers popping out of filing cabinets, they filmed paintings swinging and turning to face the wall, and heard untraceable loud sounds – and they witnessed the end of all the phenomena when Annemarie was finally discharged. They were forced in the end to conclude that the events defied any explanation they could offer in terms of conventional physics, but faced with a telephone that, untouched by human hand, kept on dialling Germany's speaking clock, had to admit that some of the events at least "seemed to be carried out under intelligent control".[204]

At the same time as these events took place in Bavaria, the Miami warehouse of a firm selling souvenir glassware was having trouble with mysterious breakages. Mugs and glasses, jars and bottles, many of them hand-painted with gaudy Florida scenes, were leaping out of store and shattering on the floor almost as fast as the debris could be swept away. The police were called in, but even when four officers were on duty and had all the employees in sight, objects continued to "scoot off" open shelves on erratic paths. William Roll, Director of the Psychical Research Foundation in North Carolina, came in to investigate and recorded 224 separate incidents of objects moving, falling or being displaced. He and Gaither Pratt, by now a full professor at the University of Virginia, took up separate watching posts and observed and made detailed measurements of movements, some as large as 22 feet, several taking place in the presence of a professional conjuror who could find no evidence of fraud.[311]

Roll realised that the phenomena centred on another nineteen-year-old (it seems to be a dangerous age), this time a Cuban boy called Julio, who seemed outwardly passive and calm, but had been separated from his mother several years earlier and was later diagnosed as someone with "feelings of unworthiness, guilt and rejection . . . and dissociated tendencies, especially in relation to expressing aggression".[309] Julio was a shipping clerk, responsible for sorting and despatching the glassware and though he was kept under close observation and was never caught faking the phenomena, only one of the 224 incidents took place in his absence. Assuming that he was somehow involved and exercising a field of mechanical force, Roll analysed the trajectory of objects in relation to Julio's position at the time they moved or shattered. He found that glassware beyond a certain distance from the boy was unaffected; but that objects close to him tended to make short, outward, clockwise movements to his right; while objects further away made longer, anticlockwise movements, at right angles to him, and to his left.[312]

Roll suggested that such effects are consistent with, and could be produced by, a cigar-shaped "rotating beam" of force, something like a phased-array radar, emanating from the boy's head. But neither he nor anyone else has ever been able to record such a novel form of energy, or link it with any known structure or process in a human body.

The Rosenheim and Miami measurements remain amongst the best yet made in any poltergeist situation, but they bring us no nearer a solution. The best that can be said about spontaneous poltergeist effects is that they are real and perennial, and that nothing appears to contradict the theory that they are produced by a living agent on or close to the scene. We are still, however, very much in the dark concerning the nature of the force or forces being used, but attempts are now beginning to be made to tame the wild phenomena and to produce "poltergeists to order" under better controlled conditions.

British psychologist Kenneth Batcheldor believes that paranormal phenomena are actually rather normal and can be elicited from almost anyone in the right circumstances. He gathers groups of "sitters" together in situations reminiscent of old-time séances and when the conditions seem right, when the groups have become suitably uninhibited, he seeds the session by cheating an unusual event. He deliberately produces a rap on the table or a movement of a piece of furniture in such a way that it appears to be due to invisible forces. And by doing so, seems to overcome both an unwillingness to accept personal responsibility for such things, and an emotional resistance to them taking place at all. Once there has been a "primer event", once all those taking part have access to a handy excuse, they apparently go on to lend their own energies to a process Batcheldor suggests is collective and additive and leads in the end to real phenomena.[12]

Batcheldor's groups have produced the levitation of heavy tables and the "gluing down" of light ones to the floor.[11] Fellow Briton Colin Brookes-Smith has successfully replicated these results, and added the refinement of instruments such as strain and height gauges and recorders to prove later that unassisted movement has taken place.[32] Groups in Canada and the United States have found that it helps to give their "scapeghost" a name and to pretend that they are actually in contact with a disembodied spirit.[273] The effects include sudden cold breezes and rapping noises with strange acoustical signatures, but in every case it is abundantly clear that those present are wholly responsible.

My experience with one such group led me to suspect that the social situation could have been conducive to releasing poltergeist-like ability in one of those taking part – little or nothing seemed to happen when that key individual was not there. But shifting membership of other successful groups implies that the phenomena are being produced by committee, and take place as long as a suitable quorum are present. All true poltergeist activity seems anyway to require an audience. I know of no example of such phenomena appearing to plague a hermit who acts as an agent for his own benefit – though such a situation could arise in cases of multiple personality.

We have arrived, I think, at a position in which it becomes necessary to recognise the possibility that forces which require a deep-seated psychological need or powerful repression to appear spontaneously under normal circumstances, can be teased into the open by widening the base of activity. By sharing responsibility with others in a group, the joint threshold for poltergeistry in all those present is lowered and some of the strange effects leak into arenas in which no one person has sufficient need in their own right to go in for covert table-turning or the sneaky destruction of crockery. I believe that this is precisely what happens in séance and in sessions with ouija boards.

Rex Stanford, a psychologist at St. Johns University in New

York State, recognises that dramatic phenomena are usually associated with crisis situations, but suggests that these represent no more than the tip of the iceberg. He believes that our smaller needs in everyday life may be served by far less conspicuous events that go almost unnoticed – as when books fall open at precisely the reference you need.[349] Everyone has such experiences and shrugs them off as useful coincidences, but Stanford has devised a way of catching such low-level activity in action. He gives his experimental subjects a very dull task to do in one room, while a random number generator runs in another. The subjects know it is there, but what they don't know is that they cannot be released from their boring tasks until the machine next door turns up a series of numbers that occur by chance only once every two or three days. But despite this limitation, he has so far found eight subjects who can consistently escape in less than 45 minutes.[350]

I like the idea that such talent could be widespread. It makes sense of some of the odd patterns in life, without isolating poltergeistry as a freakish performance. It gives the whole spectrum of action-at-a-distance a nice biological flavour, making it something of obvious and useful value, something likely to be selected for and to survive. I am less sanguine about a lot of other laboratory-based work on what has come to be called psychokinesis – the direct action of mental influence on another object or system.

Experimental studies on psychokinetic effects all involve systems which under normal circumstances behave in a random way; and are concerned with ways of making that behaviour less random, without the use of any known form of physical energy. They began in 1934 with a long series of experiments on rolling dice by J. B. Rhine and his associates at Duke University.[302] None of these are particularly impressive in their own right. The best involves a shift of no more than a few per cent, but over long trials they add up to results which become statistically significant. The most interesting feature of the tests is a tendency for the small success rate to decline even further as a series progresses. This "decline effect" is a good indication of the involvement, somewhere in the process, of human psychology. Neither dice nor random tables get bored, but people do – especially when called on to perform the same tedious task, perhaps thousands of times.

Encouraged by this evidence of a human "ghost" in the machinery, Rhine and others tightened up their patterns of analysis and found further evidence of the effect of human mood and interest on the proceedings. This was particularly evident in experiments which offered subjects something more appealing to play with than cards or dice. One delicate little test required them to try to displace soap films lifting gently from a bubble-machine. In most of the later dice and coin tossing experiments, flaws in the original protocol were eliminated by having the objects thrown by machine rather than by hand. And more emphasis was put on "placement" – recording where cubes and other objects came to rest, rather than which way up they landed.[66] Many of these experiments were successful in the sense that some subjects produced scores that could not be accounted for by chance alone, but for me the direction and intent of much of this statistical work has been brought into question by tests at the University of Pittsburgh. R. A. McConnell set up a dice-throwing machine in his biophysics laboratory there and found that it nicely reproduced the Duke University results when he himself operated it, but he also discovered that it continued to do so when he was not even in the laboratory, but in his home a mile away – and fast asleep.[261]

Speaking yet again as a biologist, I have no difficulty accepting the possibility that some people, some of the time, are able to influence things in their environment – particularly if such action serves some useful end. Poltergeist activity in the home of a disturbed child really doesn't surprise me. I can cope with the notion that a dilute form of this talent could be revealed by bringing statistical searchlights to bear on laboratory situations which have little biological significance, but at least catch the interest or challenge the ego of a subject. I suspect, however, that this talent is probably already near vanishing point in even the best laboratory studies and I have enormous problems with the exercise of such borderline ability, from a distance and by a sleeping subject. I don't dispute the results. They seem to reveal the existence of some marginal effect, but I can't help wondering just what is being measured.

Perhaps the best clue has come during the last decade from the work of Helmut Schmidt, a physicist now on the staff of a Foundation in North Carolina, who uses the most fundamental of all random systems. As radioactive substances decay, particles or rays are emitted from them at rates which are uninfluenced by temperature, pressure, electricity, magnetism or chemical change. The rate of emission is totally unpredictable and there is no way it can be fraudulently controlled. Schmidt connected such a substance to an electronic switch, producing a random event generator that ran a variety of displays, ranging from simple lights that turned on or off to complex video games – and then asked his subjects to try to influence the outcome. Several did so with conspicuous success.[327]

It is important to appreciate that in these tests, Schmidt's subjects are concentrating on a visible goal. They are not consciously aware of, and know nothing about, the radioactive substance at the heart of the machine. And it does not seem to matter whether the generator is connected to a single isotope, or to several sources of radioactive randomness. Or whether the display is influenced directly by the disintegration within, or is programmed to reflect the average result of groups of a hundred incidents of decay. No matter how much Schmidt changed the internal design of his machine, scoring rates remained the same. Which suggests that the influence involved is "goal-oriented" – it is a single-step process, more interested in ends than means. A clear indication that human aims, rather than experimental or mechanical artefacts, are behind the results obtained.[328]

This is a nice confirmation of biological involvement, bringing us a little closer to an organic theory for psychokinetic effects, but Schmidt's work has now gone on to introduce an alarming new slant. In some recent studies, he has tried disconnecting the display on the generator from its randomising source, and having it simply play back a sequence recorded on the previous day. In other words, he is asking subjects to try to influence something that has already happened. And the awful thing is that they seem to be able to do so! In a number of such tests, subjects have now produced significantly higher than chance scores, implying that today's thoughts could be causally connected to yesterday's events.[329]

Parapsychologists, who love a good theoretical wrangle, have dubbed this effect "retroactive psychokinesis" and taken it very quickly to their hearts. There was already some doubt as to whether Schmidt's earlier results were proof of psychokinesis – a direct influence on the machine; or precognition – a prior awareness of what the machine was going to do. And now there is talk of psychokinesis being both independent of time and being responsible, in that way, for a number of other phenomena.

The argument goes something like this: If I am trying to guess what's on your mind; and I get this vivid picture of a bowl of chocolate icecream; then my choice of this particular image may have nothing to do with telepathy. It could be information projected back in time to me now from me a little later, when you suddenly turn to me and say, "Gosh, I'd love a bowl of chocolate icecream." And the argument goes on to suggest that: Anyone else who heard you say so, could also be involved; and be sending messages back in time to me, just in case my own retroactive psychokinetic talents are in disarray.

This theory is wonderfully convenient and inclusive, covering things like "experimenter effects" – which bedevil much laboratory work, ending with subjects producing the results expected of them – but it worries me.

There is some support for such behaviour in liberal interpretations of quantum theory, which suggest that events do not actually exist in a hard and fast form until they have been observed. It is, they propose, the observer who selects a required state and "freezes" it. If you want a six on your die, or a head on your coin, or a non-random number – then that's what you get. If phenomena such as psychokinesis and telepathy do exist, and I am persuaded that in some form they do, then it seems that we are unlikely to find complete explanations for them in classical physics. Our inability to do so leads some scientists to the ultra-sceptical position of having to deny all facts which seem to contradict current theory. I am not one of that number, but I am nevertheless wary of a tendency to leap prematurely to conclusions which are, by present understandings, still very bizarre.

I will come to the question of analysis in the final chapter, and take refuge here for the moment in one final experiment which seems to me to help flesh out the nature of observed psychokinesis. Charles Honorton, now at the Psychophysical Research Laboratory in Princeton, explored the effect of psychological variables on attempts to influence a random number generator, and discovered that scores only became highly significant when his subjects were working closely together as a group. Individually, they failed.[168]

You have to want something very badly before it happens, but if you can get together with others – if you can invoke the sama – it might be easier to get the desired result.

Metal Bending

In November 1973, not long after the publication of *Supernature*, I was invited to join John Taylor, Professor of Mathematics at King's College in London, on the live television programme that introduced Uri Geller to Great Britain.

Taylor and I sat on either side of the young Israeli as he wreaked his usual havoc on a pile of assorted cutlery and then went on to stop the watch on my wrist, without touching it, and to bend the second hand of another, supplied by the BBC, under its glass. We were both impressed. It was hard not to be. Taylor felt that the scientific framework with which he viewed the world was "crumbling about his ears", and went on to make a serious study of the phenomena.[370] I saw Geller perform again several times over the following weeks and was inundated with mail from hundreds of people all over the British Isles who had seen the programme and found themselves, apparently as a direct result, involved with similar happenings in their own homes.

For me, this epidemic of copycat activity was even more interesting than the original events. I took the chance to follow up some of the reports I received and found that there had indeed been some sort of induction or social infection, and that people, many of them very young children, were bending knives, forks and spoons and inflicting permanent damage on pots and pans in ways that seemed inexplicable. The whole country had been turned into a gigantic version of one of Kenneth Batcheldor's "sitter" groups, a sort of vast electronic séance with Geller as the medium, in which millions of people had been given permission, by what they saw on their television screens, to do the impossible. And this disinhibition had apparently gone far enough to allow several hundred to produce similar effects.

That television programme was a milestone occasion for Geller too. He had already spent time in the United States as a guest of astronaut Edgar Mitchell, been tested at the Stanford Research Institute and done some public performing, but this was his first live appearance on a major programme anywhere. It was the start of a process that gave him global fame and made him the centre of a controversy which rages still. He has had to ride a seesaw of opinion that oscillates between the uncritical acclaim of true-believers and the arid scepticism of arch-critics such as Martin

Gardner, who specialise in developing labyrinthine scenarios to show how fraud could have been involved.[123]

The balance at the moment seems to have come down in favour of the conjurors who claim Geller as one of their own. Even John Taylor, who at least had the courage to examine the phenomena first, is now persuaded that the "Geller effect" doesn't exist. He, and other scientists who work on the paranormal, have acknowledged their susceptibility to misdirection and deliberate fraud, and begun to take the advice of professional magicians in designing experiments with more rigorous controls. This is all to the good, but the crusade against "pseudoscience" (and it has become a holy war, complete with defenders of the faith and charges of heresy) is now beginning to suppress exploration of new ideas by scientists who fear for their reputations and sources of funding. And anything which shuts down the frontiers in this way, can only be bad for science and society.

To be accused by the new Inquisition of being "soft on Geller" has become tantamount to charges of treason against the laws and sovereign state of nature. The world is flat and Uri Geller cheats. The journal *Nature* in 1978 published a technical paper by Taylor in which he argued that electromagnetism must in some way be involved.[371] But I doubt whether any study, even if it were conducted under exemplary conditions by six Nobel laureates, could be published today in almost any journal, if it concluded that Geller was bending metal in any way except by cheating. It is both natural and proper that new ideas should be resisted, and that exceptional claims should be supported by exceptional proofs; but the current unwillingness to even submit some claims to the usual scientific process at all saddens me.

I reserve judgement on Geller, because I have never examined him personally under controlled conditions. But there is no doubt in my mind about our joint ability as a species to do in private the things he appears to do on stage. I have watched a three-year-old child, while sitting on my lap, bend a long-shanked key that I wasn't strong enough to restore to its original shape. She did this after watching a video of Geller doing it – and because no one had yet explained to her that by doing so she might be violating the laws of nature, destroying our view of the world, or condemning scientists everywhere to the whims of a hostile and incomprehensible universe. I sincerely doubt that she will be able to go on doing so when she is older and wiser in the ways of our world.

There is a distressing similarity between the squabbles which have arisen over Geller in this decade and those which surrounded other alleged psychics a century ago. Our scientific tools have improved, but so have the conjuror's techniques and it seems clear that there can never be a perfect experiment. It will always be possible to find alternative, "normal" explanations for whatever phenomena are observed, and there is no defence against the sceptics' argument of last resort – that the entire proceedings could have been faked by collusion amongst those taking part.

For me, one of the most interesting aspects of the Geller effect, is that it owes little to historical precedent. There are scattered examples in the literature of poltergeists breaking the bowls off spoons, but nobody before has claimed such an original talent. It was this originality, plus our easy access to the appropriate props, that helped fire public imagination. And the very fact that the results could be faked, and frequently were, primed the paranormal pumps in ways that seem to have liberated a number of perfectly genuine performances. There are reports of events being sparked off even by acts put on by the ever-amazing Randi – after he had made it quite clear that he was a professional magician, a consummate liar and an addicted cheat.

I suspect that the talent is widespread. Julian Isaacs, of the University of Aston in Birmingham, uses a "mass screening" method for detecting metal-bending talent in large groups of people; and claims that when an audience is drawn from a sympathetic source, such as a spiritualist congregation, he finds such abilities in as many as 5 per cent of all those tested.[187]

I am not surprised, however, by the persistent elusiveness of phenomena under laboratory conditions. John Taylor complained of the "shyness effect" which made it impossible for him to actually see metal in the process of bending.[369] Experimental physicist John Hasted at the University of London succeeded in getting several children to distort thin strips of aluminium alloy into complex sculptural forms, but the best he was able to capture on video was an image of a strip bending slightly, as though moved by an invisible hand.[157] Almost everyone involved has stories of cameras that jam at the crucial moment, or troublesome tape recorders. On a poltergeist investigation in 1971, BBC reporter Dick Tracey went to interview a couple in their "haunted" home. After a five minute recording, he discovered there was nothing on the tape, but as soon as he went outside the interview played back perfectly well. Three times he returned to the house and each time the recording seemed to have disappeared.[47]

I once worked for an untroubled week with a German television team filming healers in action in the Philippines, only to have the lights fail during the crucial few-minute treatment of a patient with a known case history who had been specially flown in for the programme. There is an almost wilful elusiveness about phenomena under close scrutiny that often leaves workers with little more than lame excuses, and leads to a lot of ribald and knowing comment from their critics. I don't know why the difficulties exist, and can only suggest that it has something to do with our own unwillingness to take conscious responsibility for talents which can be frightening and are bound to set us apart from our friends.

Almost all the laboratory experiments which enjoy any success, do so as a result of some dilution or lightening of the protocol to permit those involved to relax and treat it as a game. But, as critics have been quick to point out, these procedures also make it easier for deception to be introduced. It seems almost impossible to squeeze what are essentially spontaneous talents into the clinical confines of fraud-free control. Which is why I continue to travel on my own, without cameras or recorders, simply soaking up situations and trying to get some feel for what is going on. I am intrigued, though, by a new development which could combine the rigours of the laboratory with the necessary spontaneity of the real world.

Veteran American parapsychologist William Cox has designed what he calls a "minilab", which consists of a glass tank or dome that is lidded, locked and sealed. Inside are a variety of target objects, each connected to a switching device that activates lights and a camera if it is disturbed in any way. With such a contraption installed in the basement of a family home in Missouri, Cox claims to have obtained film of rings linking and unlinking, of objects leaving sealed envelopes, of clocks running backwards and pens caught in the act of writing their own notes.[188] Julian Isaacs in England has begun similar work with modified minilabs of his own design under video surveillance, and reports unaided object movement and metal-bending.[294]

Neither study was beyond the suspicion of all fraud, but the results at least are encouraging. It should be possible to design a small, portable and totally secure version of the apparatus, which cannot be opened without destroying it. And to put this safely at the disposal of test subjects who can keep it in their homes or places of work or, best of all, wear it permanently, in order to catch poltergeist or psychokinetic effects in action at any time. This may be the best way to quantify and document phenomena which seem to be difficult to get to grips with in the clinical and restrictive environment of a normal laboratory.

On the other hand, there may be enough going on around us already to provide the necessary evidence.

In the 1920s, a young girl called Gwynne threatened to bring a woollen mill in Yorkshire to a complete standstill entirely on her own. In mills of the time, worsted yarn was produced by brushing out a fleece, drawing its fibres out and spinning these on to a number of reels or cones at the far end of a modified "spinning jenny". It was an efficient process once under way, but difficult and time-consuming to get started. From time to time, one of the threads would break or separate and have to be retied, which was impossible to do on a frame in action, but no foreman would dream of shutting down a machine until a good number of lines – they were called consecutive ends – were out of action.

On November 4th, 1924, the manager of the mill in question was summoned to a room whose three frames had all broken down simultaneously. The mechanics were called in, but could find no faults and the machines were set in motion again, but before long all 24 ends broke down on each frame, one after another. A representative from the Research Association (Woollen and Worsted) was sent for. He checked the relative humidity of the room and suggested that the machines be earthed in case there was unusual static in the air, but again the threads parted. And the failures continued until nineteen-year-old Gwynne, who had just come to work there, happened to be sent to another department – which promptly broke down.

The owner of the mill was consulted, but this solid Yorkshire-man was unconvinced until Gwynne was brought into a room in full production and asked simply to walk, with her hands in her pockets, down a "gate" between the spinning frames. Within seconds all the ends parted. She was taken into another room with a different kind of machinery spinning heavier counts of yarn, and these broke down. And finally, as Gwynne was being hurried out of the mill by the distraught owner, several more machines, new and different ones and all in perfect working order, ground to a halt as she passed by. Poor Gwynne was forbidden to go near the mill again and set to working instead as a maid in the company canteen.[263]

Ever since the first human used the first stick to lever up the first stone – and began to wonder about the magic in the stick that performed previously impossible tasks – we have enjoyed an ambivalent relationship with machinery. Traditionally, there are just six simple machines – lever, inclined plane, wheel, wedge, pulley and screw – from which all other mechanisms are constructed. There is little in this classification to cause concern, but as soon as the pieces are assembled in ways that allow them to work indirectly, making them less immediately comprehensible, the trouble begins. It is still possible for mathematicians like Norbert Wiener to describe all machines simply as "pockets of decreasing entropy", but for every kindly labour-saving device there is a golem, a robot that steals jobs, or something amorphous like Milton's two handed engine at the door that "stands ready to smite once, and smite no more".[227]

The first complex machine to find a place in most homes and lives was probably a clock or watch, and the literature is filled with reports of them falling from walls, running unaccountably slow or fast, or just stopping "never to go again" when their owner dies. The trouble continues, and not just in the vicinity of Uri Geller. Edmund Critchley, neurologist at the Royal Preston Hospital in Lancashire, became involved with it following research on people with heart problems who are unable to wear pacemakers without throwing the steady machines themselves out of step. He found that there are a large number of people who continually take their watches to the menders, only to be told there is nothing wrong with them; and many more who have given up trying to wear a timepiece at all. And he has made the surprising discovery that the owners of watches that stop and go erratically are often victims also of the neurological condition known as Parkinson's disease. The connection is obscure, but apparently the drugs which restore nervous balance in such patients, also have the effect of regulating their wristwatches.

Michael Shallis, a physicist at Oxford University, has begun working with a number of subjects who have perennial difficulty with appliances of every kind. They suffer from the mechanical equivalent of gardener's "brown thumb", going through electric toasters, steam irons and hair dryers at an alarming rate – and he too is finding that there are medical correlates. Several of his worst-afflicted subjects are prone also to epileptic attacks – just like Alin in Indonesia and Gwynne the haunted millgirl.

It would be very interesting to run a parallel study, comparing these neo-Luddites with those others who seem to have an extraordinary rapport with machines. Most communities have someone touched by "automagic", who seems to be able to cure errant machinery by application of little more than a sympathetic bench-side manner.

Inuit people, the ones we used to call Eskimos, appear to have a particularly high percentage of such adepts, despite the almost total lack of machinery in their traditional culture. A pilot ferrying supplies up to icebound radar stations in the far north of Canada once told me of how he struggled to repair a frozen engine in the shelter of an igloo belonging to an obliging seal-hunter. He failed time and again to find the fault, until his host, who had probably never seen anything more complex than a rifle, grew impatient and showed the pilot exactly what was wrong. There may be something in the physiology of such intuitive mechanics which could give us a clue to the strange influence which many of us seem to exercise over machines.

It can be substantial. During demolition of the old Palace Hotel at Southport in 1969, the contractors were alarmed when a four ton lift began to move up and down by itself, despite the fact that the entire power supply had been cut off weeks before. The Electricity Board were called in to investigate and reported: "There isn't an amp going into the place", but just as they were leaving the lobby, the lift doors slammed shut and it shot up to the second floor. To rule out hoaxers, they set the brake on the lift and removed the emergency winding handle, but the next evening the lift performed on cue for a BBC television team. The activity seemed to be connected with one young member of the demolition squad, and only ended when the cables were cut and he and the rest of his team had beaten the irrepressible machine to death with 28-pound hammers.[46]

Such antics may be incomprehensible, but they do not come as a complete surprise to anyone who has ever given a motor car a personal name, coaxed a new one into reluctant life on a cold morning, or nursed an ageing one through its umpteenth nervous breakdown. There is something about some machines that gives them what amounts almost to an individual personality, which can be gentle and cooperative or simply bloody-minded. Most of us, at some time or another, have had cause to curse a recalcitrant lawnmower or outboard engine that seems to run perfectly well for everybody else; and it is easy to succumb to the temptation to give the thing a good kick. Which can help to make one feel a little better, but just how do you go about really insulting a machine? That may sound like a daft question, but with the invention of the silicon chip, it becomes one of real relevance.[403]

A computer is an autonomous machine. One that operates for long periods of time without input, worrying over the details in some internal model. The big ones have big worries and the really powerful ones have become so complex that nobody understands them. They are now so thoroughly unpredictable that some, like the monster coordinating the defence network of the United States, is now officially described as "psychotic".[335] It has on several occasions already, without any appropriate input, decided that the country was under attack by the Soviet Union – and had to be restrained from retaliating in kind. Given the dangers inherent in such malfunction, it is hardly surprising that research has begun into all possible sources of error. Many seem to be internal, part of the computer's own complex "psychology", but some are apparently external and yet unconnected with programming.

Just a few years ago, the subject was part of transistor folklore, but all the big manufacturers and many computer users have been aware for some time, at least at an operational level, that there are individuals who have an inimical effect on the machinery. These people are usually singled out and quietly transferred to the modern equivalent of the mill canteen, but attempts are now being made to try to identify the kind of influence involved.

Robert Morris, the first incumbent of the prestigious Koestler Chair in Edinburgh, gives as his current focus of interest, "the investigation of psychic functioning in anomalous interactions between people and equipment". A concern which cannot be unrelated to the fact that he comes direct from employment as Senior Research Scientist in the School of Computer and Information Science at Syracuse University in New York State.[18]

It begins to look as though metal-bending and silicon-psychosis have much in common and cannot simply or safely be dismissed as conjuring tricks.

Chapter Eight

DESCRIPTION

There are around half a million words in the English language, but a recent statistical study of telephone speech discovered that 96 per cent of all conversation over the wires consists of just 737 different words.[107]

We are talking more, and saying less.

This retreat into stock phrases is a cause for real concern. It is not only dull and predictable, but could be warping our view of reality. Our window on the world is already restricted by the reducing valve of our senses, and it gets even further circumscribed by language. Raw experience is difficult for us to deal with. We need to explain it to ourselves by sticking on labels or verbal symbols that give it identity, clarity and definition. The view at best is only partial, but it becomes hopelessly blurred if we lack the words necessary to do it any kind of justice.

In Great Britain, those who have gone through a full sixteen years of education up to first degree level, use perhaps 5000 words in everyday speech and up to 10,000 when communicating more carefully in writing. And the membership of select bodies, such as the International Society for Philosophic Enquiry, are said to have an average vocabulary of 36,250 words. But in the end, of course, words alone are not enough. During the last century, we have added another 200,000 technical terms to the English language, with arguable benefits to the clarity of thought and expression. What really matters is how words are used and how meanings are assigned to them.

None of us are born free. To a large extent, our minds are made up for us even in the womb by a tyranny exercised by some words that trick us into thinking along certain lines. Most European languages, for instance, have adopted the strange habit of representing time in spatial terms. Things, we say, take a *long* time or a *short* time, which encourages us to think of time in linear terms, a something made up of distinct units which can be measured. We end up, as a result, treating time as a commodity which can be *wasted* or *saved*. We allow our armies to *buy* time for embattled troops, and our television companies to *sell* time to eager advertisers. And centuries of thinking in this way has now made it very difficult for us to come to terms with time which doesn't necessarily move in an orderly way, but mixes up our artificial tenses and seems, on occasion, even to go "backwards". Other languages have managed to avoid this bias. North American Hopi, for example, never uses spatial metaphors for time or gets hung up with our rigid linear interpretation and, as a result, exercises far fewer prejudices. The Hopi see time in terms of events rather than units, describing the germination of a seed or the phases of the moon; and are never surprised when things run in cycles, bringing time back to where it started. The Hopis also avoid our confusion with past and future, by putting everything into an elaborate and subtle multiplication of the present. Where an English-speaker might be content with the comment "She dances", Hopi have the choice of dozens of equally simple forms of words which distinguish between the meanings: "I am told that she dances"; "I know that she dances at this moment, even if I cannot see her doing so"; and "I remember that I see her dancing and presume that she does so still". And the beauty of this expanded present is that it includes the possibility that her dance today might influence yesterday's success in hunting or tomorrow's weather, without causing any Hopi philosophic pain.

These are real differences in language structure, which have a strong influence on attitude and understanding. "The limits of my language," said the Austrian philosopher Ludwig Wittgenstein, "mean the limits of my world."

There are purely cultural differences which give Ifugao people in the Philippines twenty separate terms for rice, or Bedouin access to hundreds of Arabic words to describe the characteristics of camels. But beyond these, there are patterns of thought which accumulate around a language and can give different cultures widely differing capacities for dealing with reality.[160]

All languages, except perhaps artificial intertongues like Esperanto, have their problems. The fourth tone of "i" in Chinese has eighty-four variations, which can change the meaning of a word in a flash from "hiccough" to "lewd". Inuit languages have single nouns with as many as 252 inflections. Chippewa uses 6000 complex verb forms. English is comparatively simple, with only 283 irregular verbs, but we who speak it are beset with problems of interpretation produced by our use of the verb "to be".

We say "the rose *is* red". And in that simple phrase we make extraordinary and misleading assumptions about the world. We give the rose a static quality, which it does not possess. We give it the illusion of permanence, which it does not have. We confuse its whole identity by speaking in absolute terms, which make no allowance for the fact that redness is not a reality, but a quality of reflected light which looks different to anyone who happens to be colour-blind; and ignores the fact that in the short time it took to describe the rose, it had already begun to change. What we intended to say was "I see the rose as red". And what we ought to be doing, in the opinion of Polish linguist Alfred Korzybski, is abolishing that troublesome little word "is", which creates most of the confusion.[54]

A start in this direction has already been made by the recent invention of "E-prime" – the English language minus the verb "to be". Being addicted myself to wild and undemonstrated conclusions such as "It is certain . . ." and "It is known that . . .", I find it extraordinarily difficult to use. In addition, E-prime is a little prim and unpoetic – it leaves Hamlet's famous soliloquy sounding very silly – but it does make everything else a great deal more precise. Try it for yourself.

The fact that most science today is written in ordinary and often ambiguous English could account for many of the difficulties we face in trying to come to terms with phenomena whose very nature puts them beyond our peculiar and restrictive syntax. Perhaps things like telepathy, which "*is* independent of distance"; and metal-bending, which "*is* impossible"; should be described as paralinguistic rather than paranormal. And perhaps one way of coming to terms with such apparent contradictions might be to look more carefully at how others describe them.

Paranormality

In 1188, a duel was arranged in the meadows outside Mons. The contestants were required to present themselves there before the ninth hour of the day. Only one turned up and, after a long wait, claimed to have won both the duel and the dispute by default. He took his claim directly to court, which decided that the point of law was clear enough, but nevertheless went into recess. They consulted clerics, who checked the position of the Sun and confirmed that the time for the service of "nones" had indeed passed. Then, and only then, was judgement given in favour of the claimant, because in the twelfth century it was still necessary, even for trained legal minds, to take advice before committing oneself to a statement concerning the time of day.[25]

It wasn't until the fourteenth century that geared mechanisms were perfected and clocks set up in every town. Until then Europe, like the rest of the world, marked time by the rhythms of sun and season. Night and day were each divided into twelve hours irrespective of season, so summer daylight "hours" were longer than the winter ones. And months were equally arbitrary. The "hunting month" lasted as long as the hunting, anything from twenty to forty days, depending on the abundance of game, not on any predetermined calendar. Time was elastic, not mechanistic, and no one was tied to a concept of life which moved only from a known past to an unforeseeable future.[57]

Anything could happen, and it frequently did, simply because it was allowed to.

This social contract can be completely unconscious. In one recent and elegant experiment, ten subjects were placed in each of two rooms. In the first room, nine were given a stimulating amphetamine and the tenth a dose of sedative barbiturate. In the other room, the experiment was reversed. The results were dramatic. In both cases, the "odd man out" behaved just like his companions. In one room the lone barbiturate taker behaved as though fatigue didn't exist, while in the other room the lone amphetamine taker fell fast asleep.[318]

All societies create their own worlds, using language and folklore to impose an arbitrary order on the complexity of the cosmos. This ordering of reality helps make sense of things by interpreting information in ways which are compatible with what is already

215

known. In Japan, Goldilocks eats the porridge and sleeps in the bed, but instead of leaving suddenly when the Three Bears appear, she apologises profusely for her rude behaviour. Most Americans interpret the saying "A rolling stone gathers no moss" to mean that one must keep moving to succeed. For them, a moss-covered stone is one that has stagnated. But in more leisurely Ireland, where moss is endemic and carries a patriotic colour, people interpret the saying to mean that one must stay put in order to have any hope of picking up the experience and the expertise that come with age and patience.

Our knowledge of all things is determined by our perception of them, and that perception is a construction based on local expectations.[65] We are just beginning to appreciate how little direct access we have to the real world. The reality of a rainforest, for instance, is perceived very differently by a logger, a poet or a Jivaro Indian. They may walk beneath the same trees, but they live in very different worlds, each of which is equally valid. There is no "correct" version.

Anthropologist Paul Radin, on the track of the most sacred myth of the Winnebago Indians, collected several very different versions and when he got impatient and remonstrated with one informant, was told "That is my way of telling the story. Others have different ways."[293] That was all. No judgement was passed or implied. Oral traditions, like beliefs and perceptions, are flexible growing things until they get fossilised in writing or set solid in scientific dogma.

Some traditional beliefs seem absurd. An eloquent Objibwa Indian once told a visiting ethnologist, "We know what the animals do, what are the needs of the beaver, the bear, the salmon, and other creatures, because long ago men married them and acquired this knowledge from their animal wives . . . Today you say we lie, but we know better."[194] Part of this knowledge is certainly empirical, the result of generations of experience that give a people fine discernment. The Hanunóo of the Philippines classify their local plant world into 1800 mutually exclusive kinds, while the best that modern botanists can do in the same area is 1300 species.[64] But there is another kind of knowing that is more intuitive.

Working with ethnobotanists in South America, I am constantly amazed not only by the depth of tribal Indian knowledge of their environment, but by the way this has accrued in a comparatively short time. Most estimates of human habitation of the Amazon

Basin suggest that it goes back 10 thousand, perhaps 15 thousand years – no more than 500 generations. Scarcely long enough, it would seem, to uncover some secrets by trial and error alone. There are herbal cures now in use by people like the Yanomamo which require a dozen intricate steps in preparation, the omission of any one of which not only renders the potion useless, but lethal. And it is hard to see how this complex procedure came to be applied to this one plant, selected from a forest containing hundreds of thousands of others, by trial and error alone. When you ask the Indians about it, they simply say "The forest told us what to do."

Well, perhaps it did. Anthropologists are learning to take what people tell them at face value. One of the bravest of the new breed is Michael Harner of the School for Social Research in New York, who not only listens, but gets involved. While living with the Conibo in the Peruvian Amazon, he was told that he would never understand them unless he took a sacred drink made from *ayahuasca*, the "soul vine". He did and had complex visions of crocodilian demons, soul boats and bird-headed people. The experience was vivid and disturbing, with intriguing parallels to descriptions in the biblical Book of Revelation. Harner was interested, but doubtful of the value of such hallucination in helping him to understand the Conibo. He assumed that the imagery was drawn from his own cultural background, until he talked to an old sorcerer about it. To his astonishment, this Amazonian shaman told him exactly what he had seen in the vision. "I was stunned. What I had experienced was already familiar to this barefoot, blind shaman. Known to him from his own explorations of the same hidden world into which I had ventured. From that moment on, I decided to learn everything I could about shamanism."[151]

Ayahuasca is well known. It is a woody vine which contains a number of alkaloids with hallucinogenic properties – one of which has been called "telepatin", because it seems to turn those around you to glass, so that you can see through their bodies and read their minds. I have tried it in Brazil and can vouch for this effect. The most interesting thing about *ayahuasca*, however, is that it has chameleon qualities. It is a door which opens on a variety of landscapes. All those using the effusion share an experience, but the experience changes from one part of the forest to another. It always seems to be relevant, providing Harner when he tried it again in Jivaro country, with "spirit helpers" in the form of butterflies, birds and monkeys that told him things he needed to know in that particular community.

I appreciate how fey this sounds, but the experience is real, the information relevant, and the tradition widespread. The word "shaman" is taken from the Tungus people of Siberia, but is now used to describe anyone anywhere who enters an altered state of consciousness to acquire knowledge. Mircea Eliade, in a classic study first published in 1951, showed how universal shamanism is.

In China, Japan, Australia, North and South America, shamanic practices show a remarkable correspondence, not just in general content but even in specific detail, all directed at maintaining the psychic and ecological equilibrium of people who may once have shared an Asian heritage, but now live in very different habitats. Why the similarity? Michael Harner suggests that it is simply because it works. "Over many thousands of years, through trial and error, people in ecological and cultural situations that were often extremely different came nonetheless to the same conclusion." That the best way to find out about things was to slip into another state of mind, and ask.[151]

A shaman is a magical athlete, a middle man between separate realities. Some take mind-altering substances to cross the boundary, some do not, relying instead on ceremonies with drums and rattles whose sounds produce equally effective changes in the central nervous system.[267] But all seem to be able to see through the filters of culture, language and sense systems to other aspects of the real world – to hidden, non-ordinary levels of reality. Australian aborigines call this "using the strong eye", and believe that it puts them in direct touch not only with other people, but with other species. Anthropologists in a hundred cultures have heard this claim, recorded it faithfully as one of that society's perceptual sets, and tacitly dismissed it as totemic identification or a flight of fancy. But I wonder.

Arthur Grimble, in amongst his delightful tales of life as a colonial administrator in the Gilbert Islands, told of a man with an apparent affinity for dolphins. On a day arranged weeks in advance, he went with the villagers of Butaritari to a beach where their hereditary "porpoise-caller" waited. The shaman went into his altered state of consciousness to "summon his friends in a dream", and before long a huge school of the marine mammals swarmed through a gap in the reef. "So slowly they came, they seemed to be hung in a trance . . . It was as if their single wish was to get to the beach."[140] I once saw an Indonesian do something similar in the Banda Sea, summoning a rare leatherback turtle into a lagoon in daylight, bringing this huge and normally very shy reptile in close enough to touch by hand.[398] And now, in another of his enchanting letters from the Pacific, William Travis gives me a further clue.

Travis is an Englishman who has spent most of the last twenty-five years looking for tuna fishing grounds and establishing small boat industries in the outer islands. He is very good at his job. He knows from experience that tuna tend to congregate in waters of a certain colour and temperature near bottom configurations that include a thousand fathom drop-off, and he has access to all the usual scientific aids – such as charts, echo-sounders and instruments for detecting thermoclines – to help him locate such conditions. But when it comes to actually finding tuna in the flesh, he relies instead on ritual:

> I go out at noon-time, single-handed and not having eaten, nor yet hung-over. I head straight away from land, not thinking of anything and *particularly* avoiding any thought of the fishery itself. I steer haphazardly, apparently at random (but possibly picking up unconscious cues such as bird movements or swirls of current). Towards dusk, I let the boat drift. I wash in several buckets of seawater, don't dry, but stand on the bows nude, facing the setting sun – and then I start to think about tuna. I imagine them in the deep. I see their colour, their shape, their movement. I try to sense the ocean around and in them. I try to immerse myself with them. And when it feels right, I start up or set sail and travel in a straight line. It can be anywhere between 100 yards and 10 miles, and then suddenly I stop. There is no apparent reason, no alarm-bells in my head – I just stop and put down the long lines. At this point I feel very tired and go back and bed down, sometimes for just a short while, but it can be six or seven hours. And as I sleep, I dream the tuna and they always strike within twenty minutes of my waking.[379]

Travis says the system never fails. Sometimes, "for no reason except a vague discomfort", he aborts a trip before reaching the stage where he has to wash and strip; but once into the ceremony, and it has to be the whole sequence of invocation, he claims 100 per cent success in finding and catching precisely those large "sashimi-grade" yellowfin and big-eye tuna he needs and envisions.

I am not convinced that Travis summons tuna to him. I doubt he believes that either, but it is clear that he believes implicitly in a sequence of actions which have now become organised into a ritual. Biologists and anthropologists both recognise ritual as a repeated pattern of behaviour "that provides individuals with some predictive grasp of their circumstances".[71] His conviction obviously plays an important role in the process, sharpening his powers of observation and helping him find the tuna with the aid of normal sensory cues. But I think there is more to it. I am impressed by the similarity of his technique to those used by some shamans to help them enter the ecstatic state in which they claim to be in communion with things around them.[228] And I suspect that Travis, like them, moves through the ritual from wholly rational, logical, left brain thought into patterns of more emotional, right brain activity that actually have a wider compass. Wide enough, perhaps, to include an awareness of the presence of tons of throbbing tuna, making the water hum with their vibrant presence many hundreds of feet below.

And if an Englishman can do this with a little luck and practice . . .

Paranthropology

An anthropologist in Melanesia watched a man behave as though he were possessed by a spirit, which announced in ringing Tikopian that a large school of fish would arrive at the village beach the following morning. They did.[112]

A psychiatrist at the University of Cape Town buried a purse wrapped in brown paper and covered the site with a flat brown stone with a smaller grey one on top of it. Nobody saw him do it. He then drove for two hours to the home of a Tembu diviner, who danced himself into a trance and described the purse, its contents and its hiding place in accurate detail.[221]

A merchant in Africa equipped eight hunters, sent them out in search of elephant and arranged to meet them at a chosen spot on a certain day. None turned up, so he went to a local seer, who promised to "open the gate of distance and travel through it". In trance, this wizard told of the death of one man by fever, and of another trampled by an elephant. He described the success of a third, who was returning with eight tusks, and the fortunes of each of the other five who, he informed the merchant, were still 200 miles away and would not return for another three months. He was right in every detail.[381]

None of these anecdotes provides the sort of evidence that science requires, but they go on accumulating on the edges of the literature, nagging away at credibility, demanding to be taken seriously. What usually happens is that they get classified as "magico-religious practices" and filed away.

Half a century ago, two pioneering anthropologists made vital studies of the Navajo in North America[208] and of the Azande in Sudan.[105] Both looked at beliefs in magic, not as something sensational or primitive, but as an integral part of cultural life. They showed how such beliefs were coherent, allowing the people holding them to explain coincidence and disaster, and giving them something they could do about misfortune other than wringing their hands and worrying. Most research since then has followed much the same sort of enlightened line, conceding that witchcraft has a logic of its own and powerful social effects, but always making the implicit assumption that it doesn't actually work.[245]

Can we be so sure?

A few anthropologists have begun to wonder, but even these still find it very difficult to separate the real insights from the exotic imagery with which these things tend to become encrusted. Care has to be taken, for a start, to allow for cultural bias. When a patient in cardiac arrest in one of our hospitals reports a near-death experience that includes walking down a long dark passage towards a brilliant white light, as many such patients do, it is important to remember that the Christian heaven is already known as the "Kingdom of Light". But fortunately there are also some cross-cultural similarities.

Brent Berlin and Paul Kay at the University of California made a survey of ninety-eight languages and discovered that there was a natural spectrum of terms used to describe colours. Several New Guinea tribes use only two colour words – those for "black" and "white". In Africa, the Tiv use three, the Ibo four, the Bushmen five and the Hausa six. Given the different cultural needs of such people, this variation is not altogether surprising, but it is fascinating to know that all languages which use three colour terms, retain "black" and "white" and add the same third colour, "red". Those that have four colour terms, retain "black", "white" and "red", and add "green". The fifth to be added is usually "yellow", the sixth "blue" and the seventh "brown" – and always in this sequence. A language never gets a word that means "brown" unless it already has words for "green", "yellow" and "blue". In other words, there is an evolutionary progression in language which means that people everywhere add words for colours to their vocabularies as they need them, but in an orderly sequence.[20]

No one has yet offered a suitable explanation for this astonishing regularity, but it is a reassuring confirmation that, no matter how much our interpretation of reality might differ, we are at least enjoying some kind of common experience. And this experience very often contradicts logic and scientific expectation.

The Australian anthropologist A. P. Elkin found that Aborigines frequently seemed to be aware of events taking place at their homes, even when these were hundreds of miles distant and they had been away from them for months. He was particularly impressed by their absolute assurance. "A man will suddenly announce one day that his father is dead, that his wife has given birth to a child, or that there is some trouble in his country. He is so sure of his facts that he would return at once if he could."[97] Elkin checked on several such examples and, finding them too often correct in substantial detail to be dismissed as chance or coincidence, decided that contact of some sort was being established. He studied Aboriginal shamans or "clever men" and concluded that deliberate communication might be possible. The vision of a shaman, he said, "is no mere hallucination. It is a mental formation visualised and externalised, which may even exist for a time independent of its creator."[98] These entities are described by Aborigines as "power animals", which can sometimes be seen by others, and are capable of themselves working or "seeing" at a distance.

Similar talents have been described for people in the Caribbean, India and North America.[382] In Africa, I have myself been given information that is unlikely to have been made available by normal means. A Kamba diviner I went to visit in Kenya on one occasion threw the bones for me and described, in detail that meant very little to him in the tropics, a boat in a winter storm at sea. It was not until I returned to Nairobi several days later that I learned of the wreck in the Antarctic of an expedition ship on which I worked for several years, and which involved several close friends.

Reports of "crisis telepathy" are common enough in our own culture – many people claim to have had direct knowledge of the sinking of the *Titanic*. If distant communication exists at all, then it is not surprising that news of great crises should be widely broadcast. But most of the best evidence continues to come from other less technical, less sceptical, cultures – and much of it suggests that crisis calls are not so much broadcast as finely focused.

One of the best examples in my own files concerns a Cajun from New Orleans, a tough thirty-two-year-old Creole called Shep, who joined the crew of a fishing boat working deep waters at the north-west end of the Hawaiian archipelago. On the evening in question, they had been trawling and, in a quiet moment, Shep decided to go to the crew quarters. He grabbed the hatch rail and swung himself, as he always did, down into the forecastle. This time, however, his hands were slippery with fish scales and he fell flat on his back on the deck below. Nobody saw the accident and Shep lay there, paralysed and in intense pain. He was convinced that he was about to die, wondered what would become of his young American friend Milly, noticed that the time was 9.12 and passed out.

On the main island 600 miles away, Milly was visiting the home of the boat's captain, passing the evening in a little social embroidery. The wife of the skipper was a full-blooded Samoan, who was intent on her needlework, chatting away cheerfully, until she felt what she later described as "a blow at the back of the head". She slipped semi-conscious to the floor and when she could speak, said "Something very bad has happened on the boat." And then added, "It isn't Bill" – her husband. When Milly looked at the clock on the wall, the time was 9.14. It was not until the early hours of the following morning that the Coast Guard called to tell her that Shep had been landed on Kauai with a broken back and was being flown home.[379]

This incident is interesting for two reasons. The first is for the evidence it apparently provides of a true crisis call, an alarm prompted by need and carrying a message of biological significance. The literature and folklore of most people include such events. The second feature is less common and more interesting. The sender was a man from a culture which, at least unconsciously, allows such things to happen. The message was intended for a woman whose upbringing made her less receptive and, when she proved unresponsive, it appears to have been re-routed to another person nearby who was only indirectly involved, but whose cultural background and perceptual set made her more sympathetic. Once again it seems that these things are goal-oriented and not only independent of distance, but indifferent to route and means. It is results that matter.

Our Western culture is probably the worst one of all in which to look for unusual talents. Even those few scientists prepared to admit the possibility of the paranormal see it as something minimal and capricious that needs mathematical massage before it becomes manifest at all. And this attitude has carried over into the few serious attempts at cross-cultural studies.[307]

In 1943, Plains Indian children at a school in Manitoba were tested for clairvoyance with the usual Zener cards developed by J. B. Rhine.[116] In 1949, a group of half-caste Australian Aborigines were tested with both cards and dice.[315] In 1953, further such tests were applied to another small group in New Guinea.[279] And in that same year anthropologists Ronald and Lyndon Rose completed a long series of similar tests with fifty tribal Aboriginal subjects in Central Australia.[316] The Roses spent enough time with the people they studied to gain some understanding of the role of magic in their lives, and their tests showed an interesting tendency for pure tribal subjects to score better than detribalised ones. They wrote a book about their experience called *Living Magic*, which contains some fascinating insights, but next to the rich account of Aboriginal belief, their laboratory-style experiments look pale and irrelevant – as indeed the cards and dice must have seemed to those being tested.[317] They also went on to do comparable work in New Zealand and on Samoa, usually getting results that could be shown to have some statistical significance, but never turning up evidence that was more convincing than anything already achieved in North Carolina.

In the 1960s, psychologist Robert van de Castle from the University of Virginia, made a three-year study of Cuna Indians in Panama. He worked with 344 subjects on the San Blas Islands and at least tried to make the tests a little more culturally relevant by replacing the sterile Zener symbols of squares and circles with more meaningful local images, such as canoes, shells and jaguars. But the results remained only marginally significant. To give him credit, van de Castle was aware of the absurdity of what he was trying to do. It is often extremely difficult, he said, for an experimenter "with his unfamiliar paraphernalia and his blunt insistence upon having some unusual phenomenon demonstrated at a time and place convenient for him . . . to secure sincere cooperation from a native practitioner".[381] True; not only difficult, but ridiculous.

I appreciate that the intention is to acquire data that will be strictly comparable with those already collected in our culture, but the whole endeavour is back-to-front and ill-conceived. What we have is a powerful force that appears to be in active use in tribal life, but fades to a barely discernible vestige in our society. And what these researchers have been attempting to do is to identify the source with techniques designed to detect microscopic traces. Playing dice with diviners is a little like trying to illuminate the Sun with a flashlight.

There has really only been one significant development in paranthropology in the last decade – and that is the adventure recounted in his series of books by Carlos Castaneda. Whether his account is fact or fiction is irrelevant, for what Castaneda has accomplished stands as an achievement in its own right. He has drawn attention to the narrowness of our conscious experience, and the deep-seated hostility he has provoked in some quarters is a measure of the success he has achieved in throwing cultural windows open to the world of other realities. The actual existence of Don Juan is neither here nor there, but his manipulation of spirit power is fascinating, and Castaneda's problems with his apprenticeship ring uncomfortably true to anyone who has ever been exposed to real magic.[49-53]

The popularity of Castaneda's books owed a great deal to the psychedelic enthusiasms of the 1970s, but they have a more enduring relevance. Each contains a number of exciting protocols for exploration of that wider reality that has come to be called "inner space". And as a body of work, they have already inspired a small number of new and unorthodox approaches to the paranormal.

Scott Rogo at the Kennedy University in California, after considering the tricks which shamans may use to fake their wonders, suggests that some may be genuinely gifted with psychic ability. "Shamans are great tricksters, but keep on the lookout for the real stuff!"[306] A number of open-minded researchers have been looking at healers who use dramatic techniques, without automatically assuming that the effects must be fraudulent.[28] When a Philippine healer produces blood that isn't human, it doesn't necessarily mean that he took it that morning from a chicken. Jerome Clark has reviewed Mormon records of encounters with angels as though they were, at least, real experiences for those involved.[62] Michael Goss has unearthed evidence to suggest that folkloric accounts of picking up a "phantom hitchhiker", might be based on a genuine phenomenon.[129] And David Barker, after an encounter with a Tibetan lama who apparently diverted a rain storm, was able to suppress his automatic scientific rejection of the event as coincidence, for long enough to admit that it might actually have happened. Though, he says, "the whole experience produced in me a feeling of distress and disorientation which persisted for weeks."[9]

I know the feeling.

There are signs that some science could be stretching its boundaries enough to encompass non-ordinary reality, without falling prey to what some super-objectivists have called "involuted, ethnocentric, irrational, and subjective modes of consciousness".[153] Gullibility isn't a necessary consequence of being open-minded – but enlightenment is.

It is time that the barricades came down. I believe that those interested in the unusual will find that it can be more easily found and better understood in the flesh, than on the printouts of polygraphs and computers.

Parapsychology

We need a few definitions:

The paranormal I take to refer to those phenomena which in some respects seem to exceed the limits of what is deemed possible, at least on current scientific grounds. In other words, anything our culture at the moment finds surprising or unlikely.

Paranthropology, if such a science were to be recognised, would be the study of such paranormality in any of the 4000 to 5000 human societies for which we have any evidence.

But parapsychology, which literally means "the rational discussion of things beyond the mind", is that study in just one culture – ours.

The roots of parapsychology go all the way back to tribal sources, but rest more recently on a kind of curiosity that was very Victorian. It coincided with a technological revolution that was marked by the death of Michael Faraday and the birth of Ernest Rutherford. The Victorian scientists and scientific naturalists, fired by a series of successes, rapidly built up a strongly materialistic and reductionist world view. This carried enormous authority, but in an era that saw electric trams, internal combustion and the discovery of radioactivity, telepathy to some seemed no more unreasonable than Edison's telegraph or Bell's telephone – and equally accessible to study, at least under hypnosis.

By the late 1870s, there was a need for a thoroughly developed, scientifically based alternative to the narrow dogmatism of the day. Psychology had just come into being as a discipline distinct from philosophy and was very jealous of its delicate reputation, but physicians felt less threatened. And it was groups of philosophers and Victorian medical men that got together, on both sides of the Atlantic, to study subjects that lay outside the boundaries of recognised science, but which seemed in their words, "to present certain points of connection amongst themselves". And they christened their study "psychical research".

The establishment didn't like it – any more than they do now. They argued for a rational and scientific interpretation of experience, and for concentration only on physical phenomena that could be known and described. All else was metaphysical and therefore pre-scientific – just as "pseudoscience" is said to be today. Any knowledge derived from subjective experience was not verifiable and therefore not acceptable; and strenuous attempts were made to keep the new researchers out of the scientific community. In 1876, William Barrett, Professor of Physics in Dublin, submitted a paper entitled "On some phenomena associated with abnormal conditions of mind" to the British Association for the Advancement of Science. It was rejected by the Biology Section and only received a hearing in the end because Alfred Russel Wallace voted to have it read as a late entry to the

Anthropology Subsection. It was never published by the Association.

The British Society for Psychical Research was formed, in self-defence, in 1882. Its interests lay largely in "thought transference" and in the hope that a scientific explanation could be found to demonstrate that subjective experience was a valid basis for an alternative world view. It grew, as historian Molly Noonan vividly describes, out of a philosophic crisis suffered by a small group of intellectuals, but it nevertheless went to great pains to be as objective as possible, printing on the first page of its Constitution, in italics:

> *To prevent misconception, it is here expressly stated that membership of the Society does not imply the acceptance of any particular explanation of the phenomena investigated, nor any belief as to the operation, on the physical world, of forces other than those recognised by physical science.*[270]

The American Society for Psychical Research was formed in 1884 to "explore the borderland of human experience". It adopted somewhat wider grounds of interest than the British Society and succeeded immediately in recruiting biologists, physicists, and astronomers; but ran foul of psychologists, who had just established their first laboratory at Johns Hopkins and were intent on keeping their young science strictly experimental. Psychotherapists, however, were less concerned with methodological purity, and more sympathetic; so psychic research in the United States became associated with studies of abnormal psychology.

Neither Society was part of a strictly scientific movement, but as a result of early scientific antipathy, both developed a bad case of "physics envy" and tried to bury scientific incredulity under a heap of facts. They are still trying to do so.

I have glossed over the first half-century of active psychic research with unseemly speed, in order to arrive at the landmark year of 1934 in which a young psychologist published a slim volume called *Extra-sensory Perception*, which threw the subject open to a new and enthusiastic audience.[299] J. B. Rhine set out to add scientific plausibility to the discoveries of psychical research. He wanted to naturalise the supernatural. And so, instead of working with recognised psychics or spirit mediums, Rhine developed simple techniques that he used on unselected subjects, mostly students, showing that telepathic and clairvoyant talents could be found even in ordinary people. And the people loved it. Polls continue to show that seven or eight out of every ten adults in our culture believe in "ESP", most of them on the basis of some personal experience.

Rhine also succeeded in establishing psychical research as an experimental science in an academic setting. He took the subject out of the séance, gave it the new and less emotional name of "parapsychology" and began awarding graduate degrees for work done under his direction in the new laboratory at Duke University. All of which became possible because of the enormous public interest which his books inspired. As a result of private donations at one point in the late 1930s, Rhine personally controlled over 10 per cent of the entire research budget of his university. Which naturally led to jealousy and hostility and to a pattern of criticism, much of it in the popular arena, that lowered the quality of dialogue about the paranormal in a way which persists to this day.[239]

Most criticism directed at parapsychological research was, and still is, concerned with either fraud or poor technique. Both are valid concerns. Fraud is perhaps the most damaging, because it implies that there are no phenomena at all. It is seen to range from honest misperception, through a tendency to be taken in by unscrupulous subjects, to conscious chicanery by all those involved. And accusations of deception extend from David Hume's dismissive conclusion that it is easier to believe in lying and cheating than in suspension of the laws of physics; to Martin Gardner's more rational but obsessive reconstructions of how effects *could* have been achieved by anything short of paranormal means. Except for tighter controls on future experiments, there are few defences against such attacks, which leave many parapsychologists wondering why they didn't go instead into something safe like chemistry, where you can produce new and interesting results without having to submit to a lie-detector test.[223]

As far as technique is concerned, the sloppy design and loose control of much of the earlier work rendered it very susceptible to criticism. This is not true of most experiments in recent years, which have been screwed down tighter than tests in virtually any other less vulnerable field. But even these rigorous controls have failed to satisfy critics, who have turned instead to their lack of repeatability. It is certainly true that the much-desired easily repeatable experiment has eluded parapsychology, although the same could probably be said of medicine and meteorology. These are, however, "normal" sciences which do not require the extraordinary proofs expected of claims for the paranormal.

If, say the critics, the phenomena are truly independent of all known laws of science, then there ought to be new rules to put in their place. So far there are none. Parapsychologists have no accepted theories and argue that mere repeatability is not especially virtuous anyway. They claim that there is replication in those studies which cover sequences of out-of-body experiences or poltergeist activity, and that patterns do emerge from these historical surveys which allow some predictions to be made. This has certainly been enough to permit other disciplines entry to the sacred fold – geology, for instance, is almost entirely observational and predictive, none of it is repeatable at all.

The fact is that there has been just enough replication in parapsychology for the discipline to have grown in half a century from one laboratory in North Carolina, to a subject replete with dozens of research centres, journals, societies, conferences, grants and educational opportunities; and yet not enough to avoid the home-grown conclusion that "even the strongest parapsychological evidence is, by common consent, unsatisfactory and defective in one or more respects."[17]

It has become clear even to the most committed parapsychologists that there can never be "sufficient evidence". There is no perfect experiment which will be beyond all criticism.

Not because all critics are necessarily ignorant and prejudiced, but because the consequences of accepting the finds are so revolutionary. The philosopher Hans Reichenbach is said to have been so shaken on hearing of Rhine's early results that he was heard to mutter, "If that is true, that is terrible, terrible. It would mean I would have to scrap everything and start again from the beginning."[209]

He needn't have worried. Another thirty years and thousands of experiments later, nothing is yet proven. Science has not been overthrown and the mainstream view of the universe remains predominantly materialistic. Rhine succeeded in turning psychic research from spontaneous and dramatic cases to the drudgery of repeated tests under controlled conditions. But neither he nor the hundreds who have followed him into the laboratory, have managed to produce results which change parapsychology's equivocal status. Hard work and high standards have won the subject reluctant acceptance as an affiliate of the American Association for the Advancement of Science, giving it the sort of position that psychology occupied a century ago. But it is still variously described as a frontier science, protoscience, parascience, deviant science, anomalous science, pathological science and a pseudoscience.

Parapsychology remains at least an immature science, because it has no basic principles and is bedevilled by a lack of consistent findings. It cannot any longer be accused of being unscientific. It goes about investigating, reporting and attempting to verify its observations in a meticulously scientific manner. Specific problems are raised, reliable data are sought, controls are instituted and fraud is rooted out from within. And yet, even after a century of research, parapsychology still has nothing solid or substantial to build on and, without this, has little chance of developing an established body of theory, or of being accepted as a mature science.

Part of the problem is semantic. By defining parapsychology simply as the study of what our culture finds inexplicable, we fail to discriminate between phenomena which are as yet unexplained by Western science, and those which are ultimately inexplicable. The distinction matters, because the former assumption makes parapsychology an interim science – something which will put itself out of business as it succeeds in finding the necessary explanations; and the latter implies that parapsychology is concerned with events that lie on the bedrock of nature, are irreducible and incapable of description or explanation in any other terms. The second assumption may be the right one, but it is a solution of last resort, one we have no right to accept until we have explored all other possibilities.

I am personally convinced of the reality of the phenomena with which parapsychology is concerned. I am disturbed by our continuing inability to come to terms with them, and suggest that it may help to rephrase the first assumption more broadly. Let us accept that we are faced with phenomena which remain to be explained and deal with the possibility that, though these may never be explained by our science as it stands, they are explicable in some terms. And let me make the further suggestion that the inability of current science to deal with the events, and the continuing inability of parapsychology to make such events acceptable to science in general, may be due to an unwillingness to allow mind its full role in the process. I am not suggesting that the phenomena are "all in the mind", but that our experience of reality, and perhaps even reality itself, is shaped and controlled by the mind – whatever it may be.[233]

If paranormality is redefined in this way – as something concerned with phenomena whose causes may include the mind of an observer – then parapsychology will have been well named. And we are indeed involved in a study of things which are "beyond the mind", in the sense that they have already passed through it.

Why else should controversy about the phenomena be so emotional? Sociologists Harry Collins and Trevor Pinch made a study of the way in which parapsychology has been treated by its critics in the mainstream scientific journals. They found straightforward statements of prejudice; pseudophilosophic arguments to the effect that parapsychology ought to be rejected simply because it conflicts with accepted knowledge; accusations of fraud without evidence to support them; attempts to discredit scientific parapsychology by association with cult and fringe activities; and emotional dismissals based only on grounds that the consequences of its acceptance would be too horrible to contemplate. And they concluded that the ordinary standards and procedure of scientific debate were being seriously violated.[63]

This irrational response is usually explained by pointing to the fact that parapsychology is anti-materialist and a danger to the very structure of science, but the physicist Henry Margenau has shown that it is impossible to identify a single scientific law actually threatened by the reality of paranormal phenomena. He points out that the Law of Conservation of Energy and Momentum has already been broken by discoveries in quantum physics, which also deal with non-locality or action-at-a-distance; and that nothing in parapsychological discovery contradicts either the Second Law of Thermodynamics or the Principle of Causality. The only contradictions that seem to exist are with our culturally accepted view of reality based on such laws.[225]

Psychologist Charles Tart at the University of California suggests that the emotional nature of the debate might be due to the fact that discoveries in parapsychology offer a more personal kind of threat, which operates through unconscious fears of the subject. He considers the possibility that our first experience of paranormal phenomena is the rapport we all enjoy with our mothers, which often seems to go beyond normal sensory contact. And that, sooner or later, all of us are faced with a conflict created by the disparity between mother's overt messages of kindliness and concern; and her very human, covert and occasional fury and frustration with us. A child in this situation *must* believe the approved social message that mother acts only in its best interest. And the response, says Tart, is a suppression of information coming in at a paranormal, extrasensory level. We learn to fear and avoid the information channel that brings in messages that create such terrible conflict, and react to all attempts to revive it with deep unconscious hostility.[366]

This is clearly not the whole answer, but I find such psychological explanations very persuasive. They seem to attend to the actual situation on this ragged fringe of science in ways that purely physical theories cannot do. They also draw attention to the possibility that science is not so much a body of undisputed fact as a set of perceptions about facts that are on the whole very useful, but occasionally harden into dogma. Normal science is guided by a substantive set of beliefs held, both consciously and unconsciously, by the scientific community. And anything that starts out by defining itself from the beginning as paranormal is likely to be dismissed as irrelevant.

For me, the greatest danger posed by the traditional experiments of formal parapsychology, the ones initiated by Rhine, is that they tend to trivialise the paranormal. I find it hard to accept that statistical conclusions say anything meaningful about the phenomena. All that emerges from mathematical procedures is an indication that the results are unlikely to have been observed by chance. They tell us nothing about the force responsible for the deviation and make no direct impact on the consciousness of anyone involved. In most experiments the shift from expected values is so small it is totally invisible until squeezed into reality later by statistical techniques. I'm not questioning the reality of the shift, simply its significance in a protocol deliberately designed to keep it almost subliminal.

As a biologist, I remain unimpressed by most of the traditional work which continues to try to demonstrate the reality of the paranormal in situations which are themselves abnormal and unlikely to release the emotional tensions which I believe are essential to the phenomena. I have more sympathy with the new generation of younger parapsychologists who are prepared to accept the phenomena, and are interested instead in finding out how they relate to physiological processes, and how they may be influenced by the psychological forces that govern the rest of our behaviour.

And I am intrigued by answers to both these problems that may be offered by recent discoveries in physics.

Paraphysics

The nice thing about the new physics is this – everything which is *not* forbidden, occurs.

Relativity theory and quantum mechanics have swept away the old concepts of space and time, and demonstrated that everyday understandings of reality have been badly misconceived. The concrete world of matter as perceived by our senses, has dissolved into ghostly patterns of energy; and mind has somehow become interwoven with fields and particles into paradoxical forms which are completely alien to normal experience.

Sir Arthur Eddington, eloquent as ever, examined the dilemma of a physicist about to enter a room. It is, as he says, a complicated business:

In the first place, I must shove against an atmosphere pressing with a force of fourteen pounds on every square inch of my body. I must make sure of landing on a plank travelling at twenty miles a second round the sun – a fraction of a second too early or too late, the plank would be miles away. I must do this whilst hanging from a round planet head outward into space, and with a wind of aether blowing at no one knows how many miles a second through every interstice of my body. The plank has no solidity of substance. To step on it is like stepping on a swarm of flies. Shall I not slip through? No, if I make the venture one of the flies hits me and gives a boost up again; I fall again and am knocked upwards by another fly; and so on. I may hope that the net result will be that I remain about steady, but if, unfortunately

I should slip through the floor or be boosted too violently up to the ceiling, the occurrence would be, not a violation of the laws of Nature, but a rare coincidence.[84]

And these are only some of the minor difficulties. It might be wiser, Eddington decides in the end, to be an ordinary man for the moment and just walk through the doorway.

Physics deals with shadows. It always did, but the new physics has become aware of the fact, while the old physics operated under the illusion that it dealt with the world itself. The change is best expressed by another knight of the new realm, Sir James Jeans:

> The outstanding achievement of twentieth-century physics is not the theory of relativity with its welding together of space and time, or the theory of quanta with its present apparent negation of the laws of causation, or the dissection of the atom with the resultant discovery that things are not what they seem; it is the general recognition that we are not yet in contact with the ultimate reality. We are still imprisoned in our cave, with our backs to the light, and can only watch the shadows on the wall.[193]

To most of us still, the shadows seem real enough. They appear to be the whole world, unless something happens to make you doubt your perceptions. Jeans and other sensitive physicists party to the complex mathematics became aware of the shadowy nature of the whole enterprise, and turned round to look at the light. They found, to their surprise, that even their new physics could tell them nothing whatsoever about the world outside the cave. To go beyond the shadows, they discovered, was to go beyond physics altogether and into metaphysics. With the result that every one of them – Einstein, Schrödinger, Heisenberg, De Broglie, Planck and Pauli included – became a mystic.[411]

The weird and abstract concepts of the new physics have a strong appeal to practising mystics, moving in the opposite direction, so that they and the physicists, having met at the mouth of the cave, have come to share a descriptive vocabulary. It is important, however, to appreciate that physics cannot explain the mystical. It was, on the contrary, the spectacular failure of physics that put them on to this common ground. But there are some implications of the new physics which provide insights for the paranormal – and these have been seized on hungrily by those intent on finding an explanation.

The central feature of quantum theory is a principle described by German physicist Werner Heisenberg in 1927. Suppose, he said, that you try to examine an electron. We can only see things by looking at them, which involves bouncing photons of light off them. A photon is too small to upset something like an elephant, so we don't expect an animal of this size to be disturbed by our gaze, especially if it is not aware of our presence. But it is a more serious matter for an electron, whose position and momentum are drastically altered. A single photon can knock an electron off orbit and out of its atom altogether. So, it is impossible to measure the position and the momentum of any electron. Not because we lack the tools or the patience to creep up on it unawares, but because it can't be done. "We cannot know," said Heisenberg, "as a matter of principle, the present in all its details."

This is the essence of the Uncertainty Principle, which challenged all the beliefs of the old physics in a universe that was causal, in which one thing led logically to another. At one stroke, it demolished everyday misconception about the connectedness of events and altered our view of the world from hard physical reality to one of statistical probability.

Imagine, suggested Austrian physicist Erwin Schrödinger, a box that contains a radioactive source, a Geiger counter, a bottle of cyanide and a live cat. Everything is arranged so that in one hour there is a fifty-fifty chance that the counter will detect an atom of radioactive decay, trigger a device that opens the bottle of poison, and kill the cat. If there is no decay in that time, nothing happens and the cat lives. There is no way to predict the outcome and no way of knowing what has happened to the cat until we open the box at the end of the test period. But, suggested Schrödinger, what if we don't open the box? What then can be said about the cat?[138]

According to quantum theory, there is an equal probability that the cat is alive and dead – and neither is true until we open the lid and it takes on one of these two states. The cat is both alive and dead at the same time and in equal proportions, until we impose our consciousness on the equation, and push it, as a photon knocks an electron, one way or the other. In other words, reality doesn't exist until we observe it. History has no meaning, the past has no existence, unless it is recorded in the present.

The new physics says, in effect, that it is impossible not to be involved. There is no such thing as an objective experiment, because there is no result at all unless an experimenter is there, observing what happens, playing a part in deciding what happens. This is, at first encounter, a weird and purely metaphysical notion, but it is an accurate description of how we actually operate. Our minds don't perceive what is out there in the world, but what we believe should be there. We decide the cat's fate.

Cyberneticist Heinz von Foerster says, "This should not come as a surprise, for indeed 'out there' there is no light and no colour, there are only electromagnetic waves; 'out there' there is no sound and no music, there are only periodic variations of the air pressure; 'out there' there is no heat and no cold, there are only moving molecules with more or less mean kinetic energy, and so on. Finally, for sure, 'out there' there is no pain."[385]

It is the brain that sees, hears and feels. We edit experience to fit our preconceptions about it. We rescue Schrödinger's cat, because that is what most of us would like to happen. We do not merely observe the physical world, we participate with it – and it is a very democratic process.

At Harvard University, students were asked to match the length of a line with one of three others presented nearby. Less than 1 per cent got it wrong. But in a second test, where the choice had to be made in the presence of a group who had been coached beforehand to unanimously choose a line that was clearly wrong, the success rate fell dramatically. Under social pressure, "wrong" judgements were made 36.8 per cent of the time, even when the length of the two allegedly equal lines differed by as much as seven inches. The experimenters bewailed the fact that "the tendency to conformity in our society is so strong that reasonably intelligent and well-meaning young people are willing to call white black."[3]

They shouldn't have been surprised. That's how things work. We are taught to conform, and nowhere more so than in the interpretation of reality. As we grow up, we learn to ignore certain aspects of reality that are considered ridiculous or hallucinatory by adults around us. We learn to see geometric forms. We reach a consensus about the existence of just three dimensions. We come to agreements about what the world should look like. There is no objective reality, and we do not just observe the physical world. We participate with it. Our senses are not separate from what is "out there", but we are involved in a complex physiological and psychological process of actually creating what is out there. The cat is neither alive nor dead until we make it so.[359]

None of this came up in classical physics. Clearly recognisable causes led to measurable effects – and that was that. Everything else was impossible. In the new physics, nothing is impossible. Some things are less likely than others, but anything can happen. Mind and matter co-exist, the state of systems depends upon those observing them, and it no longer makes sense to separate the physical world from the psychical world.

The appeal of this understanding to parapsychology is obvious. The fact that the presence of a human observer seems to bring about "the collapse of the wave function" of a system into one of several possible states (a living or a dead cat, for instance), implies the existence of a psychophysical interaction. Determined efforts are now being made to identify this mechanism, because it would explain everything from poltergeists to bent spoons.[222]

Another consequence of quantum theory is the understanding that once two systems have interacted, the future state of each is inextricably bound up with the other. The two have to be treated as one. So, if one is observed at any future time, thereby changing its state, an equivalent change will take place in the other system, no matter how far away it may be. And this change will take place instantly, in defiance of the old principle that no signal may travel at a speed greater than that of light. Which, if it can be demonstrated, will provide a mechanism for the phenomena of telepathy and saman contact.[389]

There are few physicists who doubt the conclusions of quantum mechanics. They have already proved to be too useful to be wrong. But there are some who are discomfited by its inclusion of conscious observers, and many who find themselves bewildered by the speed with which black holes, timewarps and ideas of non-locality have become the focus for cult enthusiasm and mystic adulation. A few have even expressed the hope that the present understandings are temporary and that confusion will pass when a neoclassical and more objective solution can eventually be found.

But it is not likely to happen that way. The old ways of looking at the world have given way permanently to a new and more fundamental physics and cosmology. This new view has far-reaching implications for philosophy and might be able to solve some of the problems posed by anomalous experience, but any attempt to use it now to explain the paranormal is premature. There will undoubtedly be improvements in understanding, but if recent experience is relevant, these are likely to be more rather than less mathematical and symbolic, and even further removed from the language of everyday use.

The best that paraphysics can offer to the paranormal at the moment is Sir Arthur Eddington's comment that "Something unknown is doing we don't know what."[359]

It is good to know, at least, that there is *something* there.

Chapter Nine

SYNTHESIS

Old physics reduced the world to building blocks.

New physics sees it instead as a process, as matter in a state of flux. Atoms are now understood as both waves and particles, and are perhaps best described as "poorly defined clouds", whose form is dependent on the whole environment.

It seems to be true that quantum theory offers little direct explanation for the behaviour of things at a macroscopic level, but it has provided one vital and general insight. Which is that, however we look at the world, we should bear in mind that fragments are illusory; things are surprisingly well connected, and reality can best be understood as a whole, which changes all the time.[27]

The implications of this are far-reaching.

If *reality* flows like a stream, then *knowledge* of such reality also becomes fluid, a process rather than a set of fixed truths.

And because all *knowledge* is produced, displayed, communicated and applied in *thought*; then thought too must be seen as part of the same eternal tide.

This line of reasoning is becoming very abstract, but let me take it just one step further with the additional observation that *thought* is, in essence, a response of *memory*. It consists of a repetition of some image or sensation, or it involves a combination or reorganisation of such repetition in a new and useful way.

So, in the end, *intelligence* turns out to be part of the flow. It is not grounded in cells or molecules, but drawn from the same moving stream as *reality*.

In other words, mind and matter are ultimately inseparable.

The artefacts around you – this book, the building in which you read it, all man-made objects – have passed from memory into the environment. And as you look up at such things, and think about them, they pass back into memory again, completing a cycle. Thoughts can have a concrete existence as part of this cycle. And if I should pass through the same room in the same building, sharing with you the experience of your environment, I too become part of that cycle. And our thoughts and our memories get entangled. And so it goes on, indefinitely far into the past.

There is no way of confining thought. We cannot say where it begins or ends. Everything flows together into one unbroken totality of movement which does not belong to any particular place, person or time.

Reality isn't a thing after all, it is much more like a thought.

The role of memory in all this is clearly crucial, but far from clearly understood. Psychology breaks the process of memory down into the four sub-functions of learning, retention, re-membering, and forgetting. Biology sees it in more organic terms as a single process, an organism's unwritten record of reality. Both regard it as something which passes through the brain, leaving a memory "trace", but neither can be certain where this is housed. Large parts of the brain can be destroyed without abolishing memory of past events. Suspicion at one time fell on RNA molecules as chemical memory banks, raising the awesome possibility of memory transmission through cannibalism, but this has never been satisfactorily proven. And now the new physics revives an old idea – that memory could be stored, or at least carried, in a physical form.

When talking about space and time, physicists represent the movement of particles as "world lines". Very simply, these describe the path of an object such as an electron over a period of time. In graphic form, an electron that doesn't move has a world line that runs straight up from the bottom of the page – the past, to the top – the future. An electron that changes its position, as well as being carried along by the flow of time up the page, will leave a world line that lies at an angle or zigzags through spacetime.

These diagrams are convenient ways of representing reality in geometric form, but physicists also insist that world lines have a real existence. They are actual routes followed by particles and there is nothing in mathematics that prevents them from being

241

retraced. In other words, things can go backwards as well as forwards in time – and memory could be nothing more than a peek back down a particular world line.[266]

More than that, because physics makes no distinction between particles in organic and inorganic matter, memory isn't something peculiar to brains. It can equally well be stored in a rock. As far as we know, rocks can't remember; but that doesn't mean that they don't hold a memory, or that we can never share it. In fact, there is a fair amount of evidence to suggest that this is precisely what some of us can do.

Pararchaeology

In 1941, Stanislaw Poniatowski, Professor of Ethnology at the University of Warsaw, handed a small stone to an elderly Pole. For twenty minutes, Stefan Ossowiecki felt the object, rolling it over and clasping it in his hand, then he spoke:

> I see very well, it is part of a spear . . . I see round houses, wooden, covered with grey clay, over walls of animal hide . . . People with black hair, enormous feet, large hands, low foreheads, eyes deeply set . . .

He went on for an hour, giving a detailed view of the daily life, dress, appearance and behaviour of a Palaeolithic people; including an account of their ritual use of red ochre and lime as cosmetics, and a description of a cremation ceremony. All of which was totally appropriate for a projectile point identified by the Warsaw Museum as belonging to the fifteen-thousand-year-old Magdalenian culture.[278]

Ossowiecki was murdered by the Gestapo in 1944, but he was tested further during the war years with another thirty-two assorted objects from the Museum – including stone tools, bone fishhooks and ceramic figurines. And in each case he provided vivid panoramic descriptions that read like eye-witness accounts of communities and technologies ranging from half-a-million-year-old Acheulian times, through Mousterian, Aurignacian and Neanderthal cultures, to the present day. These accounts were stimulated by objects that only experts could be sure to recognise, and were supported by further complementary detail when the same object was given to him again at a later date. Despite the fact that Ossowiecki was a chemical engineer with no conscious interest in prehistoric archaeology, his descriptions are not only consistent with what was then known about the cultures in question, but sometimes included information that has only come to light as a result of discoveries made since he died.[128]

He was not the only one with such a talent. In the early 1920s, a German physician in Mexico discovered that one of his patients, Maria Zierold, was able to do something similar under hypnosis. She was particularly good at describing the recent history of fragments of pumice stone that seemed to soak up impressions like sponges and release them to her later.[275] She was officially tested by Walter Prince, President of the American Society for Psychical Research, with a number of objects sealed in envelopes and shuffled so that even he did not know which one she would receive. Zierold never opened any of the envelopes, but nevertheless described one as a farewell message from a man on a sinking ship, and she described the man. The object was in fact a piece of paper with a message found in a bottle washed up on the Azores – and when the man's widow was later contacted, she confirmed that the description of her husband was accurate, right down to the scar over his right eyebrow. And in the most impressive demonstration of all, Zierold held a sealed envelope containing a letter sent to Prince from a clergyman friend and, once again without opening it, she gave him thirty-eight pieces of information about the sender. This was information neither contained in the letter nor known to Prince, but which he was later able to verify.[289]

More recently, the Dutch clairvoyant Gerard Croiset has demonstrated a comparable talent. In 1953, working with nothing more than a tiny fragment of bone, he described a cave, its inhabitants, its surroundings and a religious ceremony connected with it – astonishing the Dean of the University of the Witwatersrand who, having brought the relic from a cave in Lesotho, could vouch for the accuracy of the "reading".

This odd ability to "read" an object's world line like a book, has in it a suspicion of telepathy and clairvoyance but it is better and more precisely described as "psychometry" – which literally means "measurements made by the mind". The term was coined in 1893 and has since fallen into disrepute with parapsychologists who, fearing that it might be confused with "psychometrics" – which describes the general application of mathematics to psychology – prefer to talk instead of "token-object reading".

I don't think that does it justice. It is a talent that is virtually ignored by most modern researchers, which is a pity, because there is far more to it than a clinical test with a token. I find that psychometry is one of those bridging phenomena that helps to knit together a number of other loose ends in experience of the paranormal.

For a start, it provides easy experimental access to a faculty which seems to operate in the same strange void as mind and memory. We know from studies of learning, that memory seizes most readily on items that are recent, frequent or vivid. Psychometry seems to do the same, most often picking up traces of events which have just happened, or have happened very often, or which have a strong emotional connection. When psychic Eileen Garrett was once asked to "read" a box containing a cuneiform tablet, she gave a vivid and unmistakable description of the secretary who was the last person to handle it before the test.[224] But with most antique objects, it is more usual to get accounts of things like richly emotive ceremonies. And if the object is a room or a house, the dominant impression will not be of any of the everyday conversations which must have taken place there, but will concentrate on the one murder or rape with which it might be associated.

And psychometry, like telepathy, very often suffers from contamination with personal memory traces that are sucked into any exciting situation and end up embellishing otherwise accurate replays.[310] Having been "measured" by the mind, I have little doubt that the readings of psychometry are as prone to dramatisation as the voices in multiple personality. But this in no way diminishes the ability of our minds to make such readings in the first place.

William Roll of the Psychic Research Foundation in North Carolina recently made a psychometric discovery which casts grave doubt on many of the earlier card-guessing tests. He showed that there was no need to have any symbols on the cards at all. Some subjects in his tests responded to completely blank cards, able apparently to tell these apart even when the only respect in which they differed was a "feel" produced by their particular histories. A sealed pack of cards handled by one researcher, for inst~nce, was perceived to be qualitatively different from those handled by another, despite the fact that no differences were discernible to any of the normal senses – not even to those of a bloodhound. In another case, preference was shown for blank cards that had been singled out by the experimenters, on an arbitrary basis, as "special" – they were Jokers with nothing but an imaginary pedigree and purely mental markings.[308]

Unlikely as it may sound, it begins to look as though information about past events can indeed be stored in physical objects. And if this is true, then it is not impossible that memory, in its intricate dance with the environment, leaves similar traces on, or interacts with the physical world in ways which are perceptible to all of us – even if it is only a few active psychometrists who can bring these cues to conscious attention.

I suspect that we do enjoy something of the sort. It has never been put to the test, but I would be willing to bet that a majority of subjects, chosen at random and blindfold, could just by holding a ring or wristwatch in their hands, distinguish one worn constantly for years, from another identical but unworn object.

It is part of common experience too that some objects have a good "feel". They carry what amounts to a patina laid down by events with which they have become associated. Dealers in antiques, of course, exercise their own special discernment in these things, but when it comes to discriminating between superb fakes and the real things, all rely in the end on a very personal assessment of whether or not an object "feels right".

And everyone, at some time or another, has walked into a strange place for the first time and felt that it was welcoming; or gone into another and become convinced that it was filled with such "bad vibes" that nothing could induce them to spend a night there alone. As a rule, these feelings are consistent, some houses just are "wrong" for everyone who goes there, regardless of the weather or the phase of the moon. And whatever it is that is disturbing can rarely be put right by a fresh coat of paint or better taste in the choice of wallpaper. Such things seem to be beyond cosmetic remedy, but sometimes respond to the new and even more emotional overlay provided by a really good exorcism.

If inanimate objects can become encoded, then it would obviously be useful to students of history to know how to crack that code. A few archaeologists have begun to take the possibility seriously.

Norman Emerson, Professor of Archaeology at the University of Toronto, regularly uses what he calls "intuitive archaeology" on field expeditions. He has discovered a truck driver called George McMullen, who has no formal education and never reads anthropological literature, but seems to be able to "read" artefacts in the same way as Stefan Ossowiecki – providing information on the Iroquois Indians, which Emerson knows to be accurate. Taken to a potential site, McMullen "almost quivers and comes alive like a sensitive bird dog scenting his prey". He walks rapidly over the area to orient himself and then begins to describe the people who lived there – their age, their dress, their way of life and the whereabouts of their buildings. He once walked over a patch of bare ground, pacing out the perimeter of what he claimed was an Iroquois long house, while Emerson followed behind him placing survey pegs in the earth. Six weeks later, the entire structure was excavated exactly where McMullen said it would be.[101]

On another occasion, he helped archaeologist Patrick Reed who was working on a tenth-century Indian village buried beneath an overgrown field. Reed was sceptical about the claims of "psychic archaeology" and determined to put McMullen to a stern test.

> I thought I'd ask him where the stockade wall of the village had been. I was pretty sure it had one, but I hadn't been able to find it. George told me, "It's there", and traced out a line forty feet long. Twelve inches under the ground, I found the stockade remains. It scared the hell out of me.[128]

In 1974, British archaeologists completed excavation of a rich site going all the way back to pre-Roman times in the garden and orchard of a manor house at Chieveley. Such finds are not unexpected in England, but what made this one unusual is that, despite the lack of surface indications, accurate details of all structures, ditches and roads had been plotted for them beforehand by an intuitive aide.[99] In the United States, a group of psychics have been helping to find burial sites for the Missouri Archaeological Survey Project. Nobody has any idea about the mechanism involved, but this hasn't stopped archaeologists in Ecuador, Mexico, France and the Soviet Union from recruiting assistance which saves money by cutting right through the normally laborious and expensive procedures for conducting field surveys and making often unproductive trial digs. Emerson in Canada says that McMullen on his own "has given me enough advice on where to dig and what I will find, to keep me busy for a decade".[128]

If I am right about the psychometric ability being one which we all possess to some extent, then it should be possible to find evidence to show that some archaeologists, even without psychic assistance, are more successful than others for reasons which have little or nothing to do with field experience. It is. In 1871, despite manifest scepticism from the experts, businessman Heinrich Schliemann went out with little more than Homer's *Iliad* in hand and found the site of classical Troy on a hill overlooking the Dardanelles. And went on later to discover the important shaft graves at Mycenae and the fortified citadel at Tiryns. Similar intuition seems to have led J. J. Winckelmann to the treasures of Herculaneum and Pompeii in the eighteenth century. And to have brought Henry Layard in 1845 to the site of Nimrud, the old capital of Assyria, and to the discovery of Sennacherib's great library of cuneiform tablets at Nineveh. Beginning in 1899 and without previous field experience, Arthur Evans made a series of spectacular Minoan discoveries in Crete, ending with his famous excavation of the palace at Knossos.

Many of the finds of the Great Age of archaeological discovery seem to have taken place as a direct result of reliance on myth and intuition, rather than rational spadework. And the tradition continues today in Africa with the fabled "luck of the Leakeys" in Kenya and Tanzania, and the hunches of Donald Johanson on fresh ground further north in Ethiopia. "I found myself," said Johanson one day at Hadar,

> under a strong compulsion to put off paperwork and go out surveying instead. I knew I shouldn't. But at that moment Tom Gray came in and began asking where Locality 162 was. Combined with the powerful feeling that this should be a day spent hunting fossils, that decided me. The papers could wait. Gray and I drove out of camp. Two hours later we found Lucy.[195]

All other things being equal, it seems clear that the laurels in fieldwork, which is a chancy business at the best of times, go to

247

those explorers who, in addition to their undoubted expertise, have the courage to trust their instincts.

In the Chatanooga countryside of Tennessee, one of the most successful finders of Indian artefacts is an artist called Kenneth Pennington. He is an amateur archaeologist of Cherokee descent who never goes out looking without first offering a solemn prayer in which he chants: "Oh, ancient peoples now slumbering . . . guide my shovel to the truth and beauties in the old fields . . . that I might bear witness to your life and be your choice from the past." Once in the field, he finds himself overcome by what he describes as "an overwhelming urge" which tells him precisely where to dig. And the digging has led to discovery of an unusually large number of rare one-of-a-kind archaeological objects.[48]

Pennington's attitude to his ritual chant is very much like that of William Travis' invocation to tuna. There is no evidence that either ceremony has any direct connection to their documented success, but both men's belief in their procedure is strong enough to keep them doing it faithfully and to take them out on the hunt with a powerful certainty of finding what they seek. And this seems to be important.

There is, of course, another ancient ritual of a very similar sort. I am talking about the practice of divining the presence of unknown things with the aid of an instrument such as a forked twig – the art of dowsing.

Parabiology

The history of divination is long and success widespread. People keep discovering things that would seem to be beyond their grasp.

Dowsers, working on the basis of "no find, no pay", regularly locate sources of water where geological experts have failed to do so; and provide accurate information about depth, yield and purity. Every major pipeline company in the United States has one on its payroll, and there is at least one in the permanent employ of the Canadian Ministry of Agriculture. UNESCO have engaged a dowser to pursue official investigations for them. Marine divisions in Vietnam used bent coathangers to detect concealed Vietcong tunnels. Soviet scientists have made good use of dowsing techniques to pinpoint structures below ground at the Volokamsky

Monastery near Moscow, and for geological exploration from the air over Siberia.

They report a saving of 30 per cent in the amount of drilling needed to find and mine gold deposits in the Northern Caucasus. Even Alfred Wegener, the German geologist who first proposed the theory of "continental drift", was a dowser – once tracking a fault line with a pendulum from the back of a yak in the Urals.

Willow wands, hazel twigs, whalebone, copper wire, steel rods, pitchforks, walking sticks, amber beads on silken threads are pressed into service all over the world – and apparently succeed in doing everything from sexing fertilised eggs to diagnosing cancer. The problem, as far as science is concerned, is that controlled tests under laboratory conditions are seldom as successful. In experiments at the Military Engineering Experimental Establishment of the British Ministry of Defence, twenty dowsers failed to find buried metallic or plastic mines, and were unable to determine whether or not water was flowing through buried pipes.[370] But it is difficult to ignore "on the job" results by professionals going about their trade. The pharmaceutical company Hoffmann-La Roche always include a dowsing survey when setting up new factories anywhere in the world. Their spokesman explains: "We use methods which are profitable, whether they are scientifically explainable or not. The dowsing method pays off."[185]

The system works. And the difficulty of proof and lack of an identifiable force, prompt comparisons with many other elusive paranormal phenomena. Dowsers, on the whole, are a down-to-earth sort of people who are anxious not to be linked with the psychic world. "Anyone can dowse," insists veteran British dowser Tom Graves. "It's just a skill which, like any other, can be learnt with practice, awareness and a working knowledge of its basic principles and mechanics." These he sets out very clearly in the best practical guide to the subject, noting in discussion of mechanical aids that dowsing is an essentially mental skill – "The instruments I've described are tools to tell you what your hands are doing, and not much more."[133]

This sounds right. Dowsing seems to be a basic biological technique. I have had a forked twig twist in my hands with enough force to shred its bark, but I have never been able to rid myself of the conviction that its response was a magnification of my own unconscious sensitivity and had nothing to do with magic or mystery forces. An American electronics engineer measured the movement of such a twig with a strain gauge bending beam and decided that the force was externally applied.[22] But the fact remains that many dowsers use nothing more than their bare hands. In some communities they are still known as "hand-tremblers".

There is, it seems, good reason to assume that information received from the environment in some still mysterious way is being displayed in coded form by means of muscular movements. And the fact that dowsing shows "decline effects", is subject to hypnotic influence, works better when one is not concentrating on it too hard, responds to strong motivation, and can be inhibited by hostile or sceptical witnesses – all suggest that there is a strong mental component. There may well be subtle geophysical stimuli involved, but they appear to be directed through and modified by the mind *en route* to hand and wand.[189]

The strongest evidence suggestive of mental mediation in dowsing and psychometry is the ability of some diviners to work, not on site in the field or with an actual object, but on maps many miles away.

Norman Emerson in Canada once used psychics to dowse maps of the long Montreal River in Ontario and in one day they pin-pointed thirty-two Indian sites, which turned out to be good ones, much to the chagrin of his colleagues who had spent five years on an unsuccessful field survey of the same area.[102] In this instance, it is possible that Emerson's own knowledge and expectations had something to do with the result, but there are numbers of examples in the literature of dowsers working without any possible access to information on areas they know nothing about. Canadian archaeologists William Ross and William Noble of McMasters University, have both had conspicuous success in field excavations as a result of preliminary surveys made by simply holding a flint arrowhead over a map of an otherwise unexplored area, months before setting out.[320]

This knowledge-from-a-distance is as difficult to deal with in scientific terms as the action-at-a-distance of poltergeists. Neither seems to provide any possibility of electromagnetic explanation and both, as a result, have been summarily dismissed. Which is understandable, but remains irrational, because the evidence is substantial and widespread.

250

Map-dowsing is reminiscent of a talent developed earlier this century by a country priest in Switzerland. The Abbé Mermet decided that if a pendulum was sensitive to an underground river, it might also tell him something about a patient's bloodstream. It did. By the 1920s, he was giving regular demonstrations in hospitals, accurately diagnosing even patients that doctors had concealed entirely beneath sheets or behind screens. Bodies, said Mermet, emit "radiations" that produced an invisible flux in him, which was most manifest in his hands, in which he held a pendulum as a sort of indicator.

Mermet called his art "radiesthesia", and began to stretch the distance between himself and his patients until he was able to operate just as effectively from several miles away – as long as he had something belonging to the patient on which he could focus his attention, such as a photograph or a lock of hair. Before long, his interests also widened and he began to prospect for water on maps of distant areas and even, it is said, to amuse himself on winter evenings in the country, hunched over a street plan of Paris, counting the cars crossing the Seine over the distant and invisible Pont Neuf.[241]

Meanwhile, in the United States, a neurologist had made another fascinating discovery. Albert Abrams was treating a patient with a small cancer on his lip, and ended a routine general examination with percussion – tapping the abdomen and listening to the quality of the sound. He found a spot that made a dull sound, instead of the expected hollow note of a healthy stomach – but could only get the discordant response when his patient was standing up and facing west. This made little sense to Abrams, but it intrigued him.

He brought in a medical student in robust good health and tried percussing the unfortunate young man in every possible position and compass direction, with nothing but the normal hollow sound. But then he gave the student a small glass container with a sample of cancerous tissue and asked him to hold this against his face. As soon as he did, Abrams got the same dull percussive note as he had noticed with his cancer patient. Now he was excited.

Abrams broadened his inquiry by bringing in samples of tissue from patients suffering from tuberculosis, pneumonia and malaria – and found that each produced the dead sound in different parts of

251

his subject's abdomen. He had discovered, he decided, a new system of diagnosis.

The diseased tissues, Abrams reasoned, must be transmitting some kind of "radiation" which could be measured. So he built himself an old-fashioned variable resistance box and soon found that the effects of each sample could be cancelled out by introducing a different resistance into the circuit consisting of tissue – machine – human abdomen. Cancerous tissue, for instance, was neutralised by a resistance of precisely 50 ohms. The next step was the discovery that it was not necessary to have a whole tissue sample – it was possible to identify diseases purely by the reaction of his apparatus to one drop of a patient's blood. With this discovery, Abrams had his famous "black box" – and a fierce scientific and medical controversy.

The problem is that there is no known radiation coming from a sample of dried blood that could influence such a machine, and though Abrams got impressive results with his equipment, few other people could do so. He was dismissed as a quack and died discredited in 1924.[186]

A few years later, an American chiropractor took the next necessary step and replaced the cumbersome abdomen-of-a-medical-student-facing-due-west that Abrams used as his detector, with a simple rubber diaphragm. Ruth Drown also took another great stride by claiming that her modified apparatus not only diagnosed disease at a distance, but could treat it as well by the use of "radio therapy" to restore the energetic harmony of patients who did no more than provide a sample of blood or a lock of hair. The medical authorities were predictably outraged and in 1951 Drown was arrested. She was accused of fraud and jailed. Her equipment and her records were destroyed and she died of a stroke days after her eventual release at the age of seventy-two.[321]

Her technique survived, thanks largely to British engineer George de la Warr who copied, and then improved on, the Abrams-Drown machine. He produced the first standardised instruments which were bought and are still used, in one form or another, by a growing community of practitioners who call their art "radionics", and have developed an elaborate set of theories to account for what they do. They appear to heal at a distance by measuring, identifying and repairing distortions in the energy patterns of patients who they may know only from a drop of blood on a piece of blotting paper. This sample or "witness" is inserted in the radionics box, balanced into the circuit and a diagnosis made by holding a pendulum over a detector plate.[360]

So, evolution of this strange and gentle therapy has brought it directly back to the Abbé Mermet's technique. A sort of map-dowsing over a diagram of a distant patient. The "radiations" involved are no nearer being identified now than they were in 1924 and suspicion grows, even amongst practitioners, that they don't exist. The healers themselves are almost certainly more important to any success they may achieve than the increasingly elaborate devices they choose to focus on.

There is a divinatory technique still in use in West and Central Africa, which involves a wooden "rubbing board" oracle that is consulted by dragging a smaller wooden knob across it with the fingertips, while asking a question and mentally running through a series of possible responses. When the knob seems to stick to the board, you know you have the right answer in mind, you've got your diagnosis. This system is clearly analogous to the technique of finding anomalous spots on a human abdomen or rubber diaphragm – and equally clearly accomplished without benefit of electrical circuits. I suspect that it works just as well as the hardware, and for precisely the same reason. And I am surprised that no practitioner of radionics has yet had the courage to take the final step in the sequence begun by Abrams and Drown. They removed first the patient and then the abdomen of a living detector. Now even the rubber imitation abdomen has gone. All that is left is the machinery, which stands as the only remaining barrier between them and the logical admission that what they do as humans is to divine disease in others of their kind and treat it, if they can, by psychic means.

I have made this excursion into radionics because I find it instructive to do so. I am not concerned here with the validity of the therapy, but I am fascinated by the growth and popularity of a system which appears to measure something at a distance. And to do so, no matter how much the practitioners would like to be seen to practise a physical science, by purely mental, and still somewhat magical means.

I believe that psychometry and dowsing function in much the same way. And that there are connections too with telepathic and poltergeist phenomena. I suggest that all are demonstrations of the ability of mind and matter to interact in surprising ways; ways that give us unusual access to each other and to our environment.

I am well aware that this doesn't even begin to be an explanation, but I am excited by the potential of psychometry as a way of getting to grips with the paranormal. In particular, the use of psychometric talents in field archaeology, where predictions and readings can be verified, seems to me to provide just the right emotional and cultural triggers, taking vital research out of the laboratory and giving it all the primal excitement of a treasure hunt.

There is also something vaguely familiar about doing so. A sense, almost, of *déjà vu*.

Earth Magic

In its apparent path through our sky, the Sun crosses the plane of the Earth equator twice each year. These moments in spring and autumn, when the duration of day and night everywhere on the planet are equal, are the times of equinox. Times in all cultures for the celebration of planting and harvest, for ceremonies of renewal and thanksgiving, when people gather to celebrate their relationship to the cosmos.

Such gatherings were important and always took place on sacred sites, at points on Earth's surface which had been "divined" – given the stamp of divinity by a sense of something special that could be felt there. It is on these places that we made our first worshipful marks, decorating rocks and trees in ways that turned them into shrines. And it was around these earthen altars that there grew the first rough temples – circles of hewn and standing stone that orthodox archaeology identifies simply as sites of ancient ritual.

Many of the chosen sites occupy positions of advantage on hills or plateaux with long horizons, but just as many do not. All, however, lie at the focal points of an area, at places of energetic and emotional equilibrium that make them look right and feel good. This mood and balance are unmistakable and have led to a rich folklore which deals, amongst other things, with fairies, earth magic and human sacrifice; but has left science largely unmoved – until very recently.

In the early 1970s, a zoologist out hunting for the haunts of Britain's last small colonies of horseshoe bats, happened to pass one such group of standing stones on his way home in a spring dawn, and noticed something peculiar. He was equipped with an ultrasonic detector to monitor the high frequency sounds of bat navigation, and was startled to find a strong signal coming from the ancient site. It was a rapid and regular pulse unlike anything he had ever recorded before. He searched the area for signs of life, but found nothing and left just as the sun came up, with the slightly uncomfortable feeling that he had been eavesdropping on some megalithic conversation amongst the stones themselves.

The zoologist mentioned his odd experience later to the writer Paul Devereaux, who in turn passed it on to the Institute of Archaeology at Oxford where it reached the ears of a research chemist. Don Robins is one of a small group of scientists – physicists, geologists and electronic engineers – who call themselves the Dragon Project, after the active energy currents which are said to illuminate a landscape in Chinese geomancy. The group are interested in the factors which led those who built stone circles to such tremendous efforts, which clearly stretched the "macrochip" technology of their day to its limits. The Project members wondered, when they first got together, whether there was anything special about the sites, some physical factor which could be measured. An anomaly, perhaps, in Earth's electromagnetic field which produced unusual interactions with cosmic and solar radiation.

The problem they faced was which end of the spectrum to begin their explorations, and the zoologist's story was just the hint they needed. A sensitive wide-band ultrasonic detector was constructed and field-tested for the first time in 1978 on the Rollright Stones in Oxfordshire.

Rollright is not a major megalithic site like Stonehenge or Avebury, but it is little visited and much revered. It has a circle of approximately seventy-three stones called the King's Men – rumour has it that they keep moving and can't be accurately counted; traces of several other circles; the Kingstone – an isolated menhir; and a collapsed dolmen known as the Whispering Knights, whose capstone was once dragged down the hill to be used as a bridge across the stream, until, it is said, it too kept moving and had to be returned.

Robins arrived at Rollright before dawn on a foggy morning in late October that year. "I walked around the site," he said, "clutching the detector rather self-consciously, fully prepared to pretend that it was a transistor radio should I encounter a stray visitor. The detector showed a flickering, minimal background, but in the vicinity of the Kingstone I observed a rapid regular pulsing. This ultrasound effect was noticeable for some yards around the Kingstone . . . and faded soon after dawn."[305]

Encouraged by this finding, the group built an even more sensitive detector designed to exclude all possibility of radio interference and stray signals from local energy sources or geological faults. And they put it into action all year round. On streets and bridges, in gardens and woods nearby, there was never anything more than weak and random background noise. But at Rollright there was a consistent pulsing which could be measured near dawn on any day, regardless of weather conditions, and which rose to an ultrasonic screech lasting for several hours on those mornings in March and October which coincide with the feasts of equinox. And there are records of equinoctial rites being held at Rollright even in historic times.

At other times, the stones seemed almost to be creating an ultrasonic barrier. "This was the weirdest thing," says Robins:

You always have a background of ultrasound in the country – the movement of grasses, leaves rustling, even your own clothing. It all registers. But one morning, as we moved in and out of the circle monitoring the levels, suddenly we found that there was complete ultrasonic silence inside the circle. Our first thoughts were that it was an instrument malfunction. Then we walked through a gap in the stones and there was sound. Inside, silence – outside, normal background levels.[374]

Between 1979 and 1982, the Dragon Project went on to monitor similar ultrasonics at other sites and to extend their measurements to sources of radioactivity. Using a Geiger counter, they established count rate profiles around a background of 22 cycles per minute at Rollright. And found both "cold spots" of unusually low radiation near the Kingstone, and "hot spots" in some arcs of the stone circle, where the readings soared to more than twice the background level in five-minute flares of beta-radiation. "Sometimes," said one of the team, "the Geiger yielded more counts per minute than when it was placed less than a yard from a radioactive isotope."

On sites in Cornwall, which is largely granitic and has higher intrinsic readings – around 40 cycles per minute – there were also places that were well below the background level. And all of these radioactively "cool" spots were in the centre of one of that county's many stone circles.

"What we seem to have found," concludes Robins, "are refuges. The stones shield the interior of the circles from certain energy fields . . . we are not even certain which fields. They seem to exclude cosmic radiation, too, and that implies spherical shelter. It is as though the circles created holes in the landscape." A surprising conclusion, but one which received independent confirmation a short while later from engineer Charles Brooker, who took a portable magnetometer to Rollright and found a seven ring spiral of fading magnetic intensity inside the circle of King's Men. "The Ancestors," he wrote to the same journal which published the Dragon Project's preliminary findings, "knew exactly what they were doing . . . The stone circles form magnetic refuges – Stone Age Faraday Cages."[374]

The evidence suggests that the ancient structures are neither arbitrarily sited nor randomly arranged. The cyclical variation of both ultrasonics and radioactivity argue against chance fluctuations in sound or fixed sources of radioactivity. Energy seems to have been not only detected, but perhaps even manipulated in some way. Though why the megalith builders should have wanted or needed radiation shelters is another question altogether.

Prehistoric archaeology in Britain is an often frustrating business. It can be roughly divided into those periods when people lived without dying, and those when they died without seeming to have lived. Stonehenge and most of the megalithic sites belong to the latter category. We know from radiocarbon dating that most of the activity took place between 3500 and 1200 BC, but there is very little evidence to show who the builders were or how they lived. As a result, theories about them abound. The Victorians were convinced that the circles were built by Druids involved in dark practices, whereas the tendency now is to see the builders as gentle pastoralists, "devoted to astronomy, peace and ecological harmony, and living on an exemplary diet of unrelieved wholefoods".[56]

Analysis of megalithic building in the Mediterranean, France, Britain and Ireland shows sufficient uniformity to warrant the assumption that the builders, even if they were divided into local groups, shared at least some cultural traits. Most of the stone rings are carefully designed around a basic unit of length. In many this is the 2.72 feet that have come to be called one "megalithic yard". A number, such as the astonishing grid of stones at Carnac in Brittany, show very deliberate geometry and alignment; and a few exhibit what seems to be an elaborate astronomical relevance.[420]

It was a Druidic enthusiast, William Stukeley, who first drew attention in 1740 to the orientation of Stonehenge towards the point where the sun rises at the summer solstice on the longest day of the northern year.[356] He was followed in 1894 by Sir Norman Lockyer, discoverer of helium, Egyptologist and author of *The Dawn of Astronomy*, who used the present slight discrepancy in solar alignment of some monuments to calculate their dates of construction.[230] But it was astronomer Gerald Hawkins of Boston University who sparked off the modern revival of interest with a book in 1965 that correlated the alignments of Stonehenge with the seasonal positions of the sun and the moon. The designers, Hawkins implied, were not only aware of the spherical shape of the Earth and the causes of eclipses, but had built a Stone Age computer to transmit their knowledge to future generations.[158]

Astro-archaeology was enthusiastically adopted by eager amateurs and equally quickly decried by professional archaeologists, suspicious of the intrusion of astronomy in their earthbound field. The debate dragged on without clear advantage until 1967 and the publication of what has been described as a beautifully packaged "parcel bomb".[242] For many years Alexander Thom, an elderly and retired professor of engineering, had been making accurate surveys of megalithic relics all over the British Isles, which he assembled together with supporting graphs and statistics into a body of mathematics that took a while to be appreciated, but was far too substantial to ignore. Thom's work suggested that the builders, whose programme of construction reached a peak in 1850 BC, were aware of what, almost two millennia later, was to become Pythagorean geometry. And that they used this knowledge to create rock calendars sensitive enough even to detect irregularities in the moon's orbit caused by the elliptical attraction of the sun.[373]

In the two decades since Thom's "bomb" with its delayed action fuse, dozens of other authors have had time to digest his statistics and to enter the lists on both sides of the argument, with complex mathematics of their own brought into play to defend and refute the ancient achievement. It is still too soon to reach consensus, but it seems to be generally agreed that orientation to the more dramatic seasonal events such as the solstice and the equinox, while sometimes imprecise, is too good to have been caused by chance. And that the fastidious arrangement of stones that seem to plot lunar cycles cannot easily be explained by any other hypothesis. The evidence for stellar alignments is less clear, but even the critics concede that some appear to follow the declination of bright stars such as Rigel. The best recent synthesis is that of astronomer Douglas Heggie, who concludes that while much of the evidence is equivocal, "What we do know about megalithic science . . . is intriguing enough."[162]

Nobody questions the effort or skill involved in moving stones, some weighing more than fifty tons, over many miles of rough terrain. It was not a passing fancy or the reflection of an abstract desire for astronomical information. It was a matter of great and long-lasting concern, a reflection of need which led people to identify and adorn places believed to be centres of power and protection.

These people were not gods or Atlanteans. They were small-scale subsistence farmers who probably lived in peasant communities and led relatively humdrum lives. But four thousand years ago they nevertheless went to the trouble of setting up ritual centres at carefully selected sites, and then surrounded these with whole landscapes shaped and managed to reflect their beliefs.

In 1921, an English provincial merchant called Alfred Watkins noticed that many ancient sacred sites seemed to be arranged in straight lines which coincided with existing roads, or with traces of former ones. He began to plot the positions of stones, circles, cairns, holy wells, sacred trees, chapels, churches and traditional meeting places – and decided that these were not arranged randomly, but lay on a pattern of what he called "ley" lines.[395] Watkins died in 1935, but his old straight tracks now deface countless Ordnance Survey maps carried about the country by enthusiasts convinced of the existence of prehistoric "national grids" of earth energy. Careful statistical study has shown that many of these leys are illusory. Sites align by chance more often than one would expect. But some of the alignments are beyond dispute and clearly link places still recognised as sacred, some now surmounted by cathedrals, with ancient barrows and mounds, and beyond them to natural features such as prominent peaks or distant islands.[80]

Similar straight lines criss-cross the desert near Nazca in Peru and the *altiplano* of Western Bolivia. Many are more than 20 miles long and range from hilltop to hilltop or straddle plains marked only by Indian stone cairns or Spanish churches. These lines are understood locally as tracks connecting holy sites. The Indians call them "spirit paths". Nobody knows who made them, but they have much in common with established ley lines in Britain, and suggest that the habit of making such patterns was once worldwide.

The discovery, by scientists of the Dragon Project, that stone circles and menhirs are associated with spots of energetic anomaly, also lends credence to popular belief that subtle energy of some kind flows along earth arteries; and that it is possible for us to enhance this process by a sort of spiritual engineering that enriches rather than violates a landscape.

Salisbury Plain around Stonehenge is decorated with more than 400 barrow mounds; with arcs of ditch and moat that involved the excavation of a quarter of a million tons of chalk; with dozens of ancillary temples and shrines; with hundreds of earthworks and banks; and a broad *cursus* or avenue that runs for over two miles across the downs – and none of this monumental activity can be understood in purely functional terms. It has to be seen either as a grand folly, or as a meaningful attempt to deal with the world as the builders perceived it.

And given the now established fact that electromagnetic and mechanical forces behave strangely at some prehistoric sites, it would be presumptuous of us to dismiss the possibility that those who built them were unaware of the oddity, or were insensitive to the changes their labours induced in the natural balance.

Earth Mind

The three great Laws of Thermodynamics can be paraphrased as:

You can't win;
You must lose;
And there's no way out of the game.

Life, however, hedges its bets. It has managed somehow to find a way of bending the rules without spoiling the game. And that is precisely what good ecology is all about.

Gaia is, and Gaia works. But does Gaia know? I am beginning to suspect so.

The chances are that there always was some kind of unconscious sensitivity, a sort of saman sympathy that led the megalith builders to what amounts to nicely judged acupuncture of the Earth. They followed their instincts and contrived a culture which was relevant, both to the planet and to its environment. What they did made cosmic sense, even if it was unconscious.

I believe we now find ourselves in a very different situation. Our technology has come between us and the Earth. We have lost direct touch. We spend our city lives in suspense, perched on plastic, propped up by scaffolds of glass and steel, being sapped – like Antaeus – of natural energy. But we have an advantage that didn't exist four thousand years ago.

I think that everything changed dramatically less than a human generation ago. On the day when, like a monkey with a mirror, we saw ourselves for the first time in those shattering photographs of the Whole Earth. Something happened then, something synaptic, making us a conscious part of the Gaian nervous system. It was the day that Earth became self-aware.

The change is not unlike that which takes place during the early development of the human brain. Starting about fifty days after conception, the embryonic brain enjoys a population explosion. The number of cells in it increase by many millions every day, flowering in just five further weeks to something like the 10,000 million it will eventually contain. Then the proliferation stops, almost as suddenly as it began, and a different process takes over.

The isolated nerve cells begin to make contact with each other, reaching out with networks of fibres so that each cell communicates directly with thousands of others. Some in the budding cortex may connect with as many as a quarter of a million of their neighbours. And this tangle grows more and more complex, knitting itself into greater and greater cohesion through the rest of pregnancy and on into the first five years of life.

Human life has multiplied in much the same way. The total world population in the time of the megalith builders may have been 100 million – less than the present population of Brazil. It took 5000 years to double that number, but just 1500 to double again to around 500 million by the end of the Middle Ages. We reached 1000 million in about 1840, 2000 million in 1930, 4000 million in 1976 – and the latest projections suggest that numbers will eventually stabilise around 10,000 million in the year 2095.

Meanwhile, as proliferation moves towards its target, the process of interconnection has begun. In 1976, world telecommunications consisted of perhaps 200 million rather isolated telephones – a network no more complex than a region of the brain less than the size of a pea. But today there are over 1000 million telephones and several million telex machines, all directly connected to each other and to vast banks of information stored in computers. Our data-processing capacity is doubling every two years and, if this rate of increase is sustained, the global network could rival the brain in complexity in a very short while – perhaps by the end of the century.

So, before long, there will be as many of us, as well connected, as the cells in a brain. A sort of global intelligence.

Before 1900, it would have been almost impossible to take a planetary EEG. The only marked electromagnetic activity was an occasional spark of lightning. But it was in that year that human speech was first transmitted by radio waves, and now the aether around us hums with action on every available wavelength. The planetary brain crackles with conversation and the endless chatter of computers exchanging information. No galactic probe coming close enough now to pick up the pattern, would be left in any doubt about our vital signs. The third planet from this sun is alive and well, and getting restless. Perhaps even on the point of becoming conscious.

Psychologist Peter Russell is impressed by the fact that there are approximately 10,000 million atoms in a living cell; 10,000 million cells in a human brain; a potentially stable human population of 10,000 million brains; and an estimated 10,000 million other planets in our galaxy capable of supporting life. He suggests that 10^{10} – a nice round ten British billion – is a sort of magic number, an organisational critical mass that marks the threshold of quantum leaps in evolution.[322]

If he is right, we could be on the brink of the next big step. Seen from space somewhere outside our solar system, Gaia must look a little like an organism beginning to sense its surroundings. Or a nerve cell putting out tentative fibres, extending its nervous system in search of others similarly occupied. The beginning of the connective phase in the formation of a galactic super-organism. Which in time, perhaps, is just one ten billionth part of what science fiction writer Olaf Stapledon calls "This final creature, the ultimate cosmos, absolute spirit itself."[351]

This Russian-doll version of the universe makes an elegant organic kind of sense. I suspect that the secrets of our little paranormal mysteries lie somewhere in the cohesion that comes about in the spaces between the dolls representing the organism and the society. And that we will find the answers by looking not at the bodies of either, but in the energies they create between them. But the doll analogy is not very helpful in itself as a description of what is going on in our complex ecology. I touch on it here near the end only because it is a timely reminder of our ultimate lack of substance. We are exactly like a galaxy in our fine anatomy. Matter moves through you and me as easily as the wind blows through the branches of a tree. And the boundaries we draw at the limits of our skin are as arbitrary as those which separate our solar system from the next one.

Everything is indeed connected to everything else, in the best traditions of ecology, but it goes further than that. Everything is everything else. There is no difference – and nothing is impossible.

It was once fashionable to deplore the views of Heraclitus, the ancient Greek whose apparent pessimism led to his being called the "weeping philosopher". But in the light of the new physics, he is beginning to look quite prescient. His suggestion that even the sun was made fresh each morning may have been excessive, but he was right about the impermanence of matter and the certainty of change. And wise in his warning that:

If we do not expect the unexpected,
we will never find it.

I, for one, intend to keep looking.

CONCLUSION

Something strange is going on.

We live in a world whose realities are defined by science, which tells us how things work. And yet there are some things which don't seem to work that way at all. Our science tells us that these things are impossible and don't exist, yet they stubbornly refuse to go away. There are relatively few of them and they are often elusive and hard to control, but they are there for anyone to see. They exist. And by their very existence, no matter how tenuous this might be, they present a problem.

Some students of the unusual feel that the fact of this existence turns science on its head. Some scientists seem to agree. They find the whole possibility so alarming that, rather than have science submit to such indignity, they choose to turn themselves upside down instead. The gesture is heroic, but the posture is ridiculous.

Consider just one example.

A dowser, who claims to be able to find underground water and buried minerals with the aid of a pendulum, is tested in Wales. He walks across a valley floor and hammers in a line of stakes to show where he believes a stream to be, giving an estimate of its depth and flow. He is being filmed by a television team for a programme on the paranormal. The interviewer objects that such claims are hard to check and asks for a diagnosis that can be verified. The dowser holds his pendulum over the man to assess his state of health and makes the surprising claim that the interviewer seems to be healthy enough, except for a piece of metal in his thigh. Everyone is very impressed. The diagnosis is unusual, but it just happens to be true. The interviewer once had an operation that required the reinforcement of his femur with a metal brace. It's not something that he talks about, and the scientist assessing this demonstration concedes that it would have been difficult for the dowser to have discovered the fact about the embedded metal beforehand, without having a very efficient spy network. He notes that there was no defect in the interviewer's limb movement caused by the metal insertion.

So what happens? Is there a serious discussion about the possibility of dowsing having any scientific validity? Does he begin to wonder about organic metal-detection? No. Faced with the paradox, this distinguished physicist stands on his head. "I regard this," he says, "as a coincidence."[370] And closes the subject.

It is of course, a scientist's duty to consider all the angles. Given the extraordinary nature of the dowser's claim, he has every right – as long as all other things are equal – to favour explanations consonant with the orthodox scientific view of how things work. All things, however, are not equal here. Coincidence remains one possible explanation of what happened, but its probability is very low. And there is a point where "normal" explanations become more implausible and more far-fetched than a frank acceptance of the facts.

The fact is that unusual things do sometimes happen. I have seen them happening often enough now to be certain of that. And, as a scientist myself, I admit that they present us with a problem. But it is not insoluble and it does not require any desperate mental gymnastics. I see it, in truth, as more of a paradox than a problem. An apparent contradiction produced by poor definition rather than faulty procedure.

Science decides what is possible by reference to its definition of reality. Anything which fits the definition, is acceptable. Anything which doesn't fit is impossible and must be rejected. And the problem is that the facts of dowsing or poltergeist phenomena stand in direct contradiction to the current definition. So the issue is reduced to a choice between rival facts. The normal versus the paranormal. And, of course, the normal wins – even if it does have to stand on its head to do so.

Such contortions ought to make us suspicious of the premises that made them necessary. There has to be a flaw somewhere in the argument. And there is. What is being ignored is the point that our definition of reality is a theory, not a fact. We don't know exactly how things work. All we have is a reasonably good hypothesis. And it never was a matter of choosing between rival sets of facts. The debate concerns a set of discordant facts and their relationship to a theory of how things happen. All that is at stake is the validity of a working hypothesis. And all that is necessary to reconcile the new facts with the old theory, is an admission that the theory might be incomplete. There is no need for anyone to stand on their heads. There is no assault on the laws of nature or the principles of science, and no need for protectors of the faith or charges of heresy.

What we need is a slightly broader definition of reality. One which includes the possibility of certain things happening when humans are involved. A definition that is not so exclusive; one less inclined to dismiss certain things as impossible, and better able to deal with what actually happens in terms of probability rather than outright and unreasonable denial.

I don't have such a definition to offer. I think it is probably still too soon to frame one that will work. What we need are more facts on which to base our discussion. And that is what I have been trying to provide.

I am a biologist and inclined to think in certain organic ways, which means exercising my own kind of bias that needs to be allowed for. I am also an enthusiast and inclined to get excited by ideas, going shooting off at unexpected angles, sometimes without good and sufficient reason. And that too should be taken into account. But even so, I believe that most of what I have tried to say here makes universal sense. I have deliberately avoided using the specialised vocabulary and the ugly acronyms that make much of parapsychological literature so difficult to read. And I have tried, wherever possible, to avoid fracture of the phenomena into mutually exclusive bits, which seems to me to be a direct and misleading consequence of the adoption by a precarious new science of its own defensive jargon.

I am reluctant, for instance, to separate poltergeist phenomena from those described as psychokinetic – because I suspect that a poltergeist is nothing more than an undomesticated version of the tame laboratory dice-nudger, doing its thing out there where it belongs, in the emotional wilds of everyday life. And I have similar difficulties with the distinctions drawn between telepathy, pre-cognition and clairvoyance – which seem to me to be identical and

266

logical consequences of a timeless connection that develops between organisms at a certain level of awareness. I suspect that it is only by looking at these things holistically, as patterns that emerge together during evolution, that we will begin to be able to understand how and why the totality in biology is always and wondrously so much more than the sum of its parts.

When I wrote *Supernature* fifteen years ago, I was flushed with the discovery of so many loose ends and delighted to find that many of them, with just a little care, fitted beautifully together. "Life," I concluded then, "survives in the chaos of the cosmos by picking order out of the winds . . . Nothing happens in isolation. We breathe and bleed, we laugh and cry, we crash and die in time with cosmic cues." Indeed we do. I see no reason to question that now and have here added more strands to the web on which we hang our existence. But I find that is no longer enough. I need to make the whole process a little more personal.

In thinking about this and wondering what conclusion others have come to, I find, not for the first time, that the intuitive psychologist William James has already been there. "Our normal waking consciousness," he said,

> is but one special type of consciousness, whilst all about it, parted from it by the flimsiest of screens, there lie potential forms of consciousness entirely different . . . No account of the universe in its totality can be final which leaves these other forms of consciousness quite disregarded . . . They forbid a premature closing of our accounts with reality.[190]

My account is open. I don't have the answers, and I lack the spiritual maturity necessary to find them. I am suspicious, anyway, of "enlightenment" and the easy answers of the born-again. But, speaking once again as a biologist, I am aware at times of a kind of consciousness that is timeless, unlimited by space or by the confines of my own identity; in which I perceive things very clearly, and am able to acquire information almost by a process of osmosis. I find myself, in this state, with knowledge that comes directly from being part of something very much larger, a global ecology of mind. And while the concept is mystical and the words I reach for never quite adequate to the task, there is nothing other-worldly about the condition. It is very well earthed, and only comes at times when I am steeped in some natural cycle – bathed by a spring tide or swept along on the energies of an equinox.

The experience is literally wonderful, and leaves me with a feeling of close identification with nature, rather than a transcendence of it. It is like coming home to an enthusiastic welcome. Back to the mysteries inherent in the hidden side of things; to the entire spectrum of possibilities; everything normal and paranormal, all that is usual and unusual. A return to the whole extraordinary experience that I have called Supernature.

BIBLIOGRAPHY

1. ALERSTAM, T. & HOGSTEDT. "Birds are born with a magnetic compass", *New Scientist*: 1 March 1984.
2. ARDREY, R. Introduction to MARAIS (Ref. 235).
3. ASCH, S. E. "Opinions and social pressure", *Scientific American* 193: 5, 1955.
4. BACH, R. *Jonathan Livingston Seagull*. Pan: London, 1973.
5. BACKSTER, C. "Evidence of a primary perception in plant life", *International Journal of Parapsychology* 10: 4, 1968.
6. BADDELEY, A. D. *The Psychology of Memory*. Harper & Row: New York, 1976.
7. BAINBRIDGE, J. *The Super-Americans*. Gollancz: London, 1962.
8. BARASH, D. *Sociobiology*. Harper & Row: New York, 1979.
9. BARKER, D. R. "Psi phenomena in Tibetan culture", in *Research in Parapsychology*. Scarecrow Press: Metuchen, New Jersey, 1979.
10. BASMAJIAN, J. V. "Control of individual motor units", *American Journal of Physical Medicine* 46: 1427, 1967.
11. BATCHELDOR, K. J. "Report on a case of table levitation . . .", *Journal of the Society for Psychical Research* 43:339, 1966.
12. BATCHELDOR, K. J. "PK in sitter groups", *Psychoenergetic Systems* 3: 77, 1979.
13. BATESON, P. "Preference for cousins in Japanese quail", *Nature* 295: 236, 1982.
14. BECKER, R. O. & MARINO, A. A. *Electromagnetism and Life*. New York State University: Albany, 1982.
15. BECKER, R. O. & SELDEN, G. *The Body Electric*. Morrow: New York, 1985.
16. BEGBIE, H. *On the Side of the Angels*. London, 1915.
17. BELOFF, J. "Three open questions", *Parapsychology Review* 14: 1, 1983.

18. BELOFF, J. "The Koestler Chair of Parapsychology", *Journal of the Society for Psychical Research* 53: 196, 1985.

19. BENDER, H. "New developments in poltergeist research", *Proceedings of the Parapsychological Association* 6: 81, 1969.

20. BERLIN, B. & KAY, P. *Basic Color Terms*. University of California Press: Berkeley, 1969.

21. BERNSTEIN, J. *Einstein*. Viking: New York, 1976.

22. BIRD, C. *Divining*. McDonald James: London, 1979.

23. BLACK, M. "Brain flash", *Science Digest*: August 1985.

24. BLACKMORE, S. J. *Beyond the Body*. Heinemann: London, 1982.

25. BLOCH, M. *Feudal Society*. Macmillan: London, 1962.

26. BOHM, D. *Quantum Theory*. Prentice-Hall: New York, 1951.

27. BOHM, D. *Wholeness and the Implicate Order*. Routledge & Kegan Paul: London, 1980.

28. BONEWITS, P. E. I. *Real Magic*. Creative Arts: Berkeley, 1979.

29. BONNER, J. T. *Cells and Societies*. Princeton University Press: New Jersey, 1965.

30. BONNER, J. T. *The Cellular Slime Molds*. Princeton University Press: New Jersey, 1967.

31. BORLAND, H. "Sundial of the Seasons". 1964.

32. BROOKES-SMITH, C. "Data-tape recorded experimental PK . . .", *Journal of the Society for Psychical Research* 47: 69, 1973.

33. BROWN, B. D. *New Mind, New Body*. Hodder & Stoughton: London, 1974.

34. BROWN, R. *Immortals by my Side*. Henry Regenery: Chicago, 1975.

35. BRUNER, J. *Beyond the Information Given*. Norton: New York, 1973.

36. BRUNER, J. & GOODMAN, C. C. "Value and need as organic factors in perception", *Journal of Abnormal and Social Psychology* 42: 33, 1947.

37. BUDZYNSKI, T. et al. "Feedback-induced muscle relaxation", *Journal of Behavioral Therapeutics and Experimental Psychology* 1: 205, 1970.

38. BURR, H. S. "Moon madness", *Yale Journal of Biology and Medicine* 16: 249, 1944.

39. BURR, H. S. "Tree potentials", *Yale Journal of Biology and Medicine* 19: 311, 1947.

40. BURR, H. S. *Blueprint for Immortality*. Spearman: London, 1972.

41. BURR, H. S. & LANE, C. T. "Electrical characteristics of living systems", *Yale Journal of Biology and Medicine* 8: 31, 1935.

42. BYKOV, K. M. & GANTT, W. H. *The Cerebral Cortex and the Internal Organs*. Chemical Publishing: New York, 1957.

43. CADE, C. M. & COXHEAD, N. *The Awakened Mind*. Wildwood: Hoynslow, 1979.
44. CANETTI, E. *Crowds and Power*. Gollancz: London, 1962.
45. CARPENTER, E. *They Became What They Beheld*. Ballantine: New York, 1970.
46. CASSIRER, M. et al. "Directory of spontaneous phenomena", *Journal of Paraphysics* 3: 183, 1969.
47. CASSIRER, M. et al. "Directory of spontaneous phenomena", *Journal of Paraphysics* 5: 203, 1971.
48. CASSLIN, E. B. "Man who digs up rare Indian relics . . .", *National Enquirer*: 15 April 1973.
49. CASTANEDA, C. *The Teachings of Don Juan*. University of California Press: Berkeley, 1968.
50. CASTANEDA, C. *A Separate Reality*. Simon & Schuster: New York, 1971.
51. CASTANEDA, C. *Journey to Ixtlan*. Simon & Schuster: New York, 1972.
52. CASTANEDA, C. *Tales of Power*. Simon & Schuster: New York, 1974.
53. CASTANEDA, C. *Second Ring of Power*. Simon & Schuster: New York, 1978.
54. CHASE, S. *The Tyranny of Words*. Methuen: London, 1943.
55. CHERFAS, J. "When is a tree more than a tree?", *New Scientist*: 20 June 1985.
56. CHIPPINDALE, C. *Stonehenge Complete*. Thames & Hudson: London, 1984.
57. CLARKE, R. & HINDLEY, G. *The Challenge of the Primitives*. Jonathan Cape: London, 1975.
58. COHEN, J. *Chance, Skill and Luck*. Penguin: Harmondsworth, 1960.
59. COHEN, J. *Behaviour in Uncertainty*. Basic Books: New York, 1964.
60. COHEN, J. & PRESTON, B. *Causes and Prevention of Road Accidents*. Faber: London, 1968.
61. COLBERT, E. H. *Evolution of the Vertebrates*. Wiley: New York, 1966.
62. COLEMAN, L. *Mysterious America*. Faber: Boston, 1983.
63. COLLINS, H. & PINCH, T. "The construction of the paranormal", in WALLIS (Ref. 394).
64. CONKLIN, H. C. "The Relation of Hanunoo Culture to the Plant World". Doctoral thesis at Yale University, 1954.
65. CONNOR, J. W. "Misperception, folk belief and the occult". *Skeptical Inquirer* 8: 344, 1984.
66. COX, W. E. "The effect of PK on the placement of falling objects", *Journal of Parapsychology* 15: 40, 1951.

67. CRABTREE, A. *Multiple Man*. Holt, Rinehart & Winston: London, 1985.
68. CRESSWELL, W. L. & FROGGATT, P. *The Causation of Bus Drivers' Accidents*. Oxford University Press: London, 1964.
69. CROOKALL, R. *Out-of-the-Body Experiences*. University Books: New York, 1970.
70. DALE, L. A. et al. "A selection of cases from a recent survey of spontaneous ESP phenomena", *Journal of the American Society for Psychical Research* 56: 3, 1962.
71. D'AQUILI, E. G. et al. *The Spectrum of Ritual*. Columbia University Press: New York, 1979.
72. DARLINGTON, C. D. In Preface to *On The Origin of Species*. John Murray: London, 1950.
73. DARWIN, C. *The Origin of Species*. London, 1859.
74. DAUSSET, J. "The major histocompatibility complex in man", *Science* 213: 1469, 1981.
75. DAVID, F. N. *Games, Gods and Gambling*. Collins: London, 1962.
76. DAVIES, P. C. W. *The Accidental Universe*. University Press: Cambridge, 1982.
77. DAVIS, W. *The Serpent and the Rainbow*. Simon & Schuster: New York, 1985.
78. DAVISON, G. W. H. "The eyes have it", *Animal Behaviour* 31: 1037, 1983.
79. DAWKINS, R. *The Selfish Gene*. Oxford University Press: London, 1976.
80. DEVEREAUX, P. & FORREST, R. "Straight lines on an ancient landscape", *New Scientist*: 23 December 1982.
81. DICKINSON, G. L. 'A case of emergence of a latent memory under hypnosis", *Proceedings of the Society for Psychical Research* 25: 455, 1911.
82. DIRAC, P. *Directions in Physics*. Wiley: London, 1978.
83. DOTY, R. L. et al. "Communication of gender from human breath odors", *Hormones and Behavior* 16: 13, 1982.
84. EDDINGTON, A. S. *Science and Unseen World*. Macmillan: New York, 1929.
85. EDDINGTON, A. S. *The Nature of the Physical World*. Dent: London, 1935,
86. EHRENWALD, J. *Neurosis in the Family*. Harper & Row: New York, 1963.
87. EHRENWALD, J. "Therapeutic applications", in KRIPPNER (Ref. 216).
88. EHRENWALD, J. "Psi phenomena, hemispheric dominance . . .", in SHAPIN & COLY (Ref. 336).

272

89. EIGEN, M. "Molecular self-organization in the early . . .", *Quarterly Reviews in Biophysics* 4: 149, 1971.
90. EIGEN, M. & WINKLER, R. *Laws of the Game*. Allen Lane: London.
91. EINSTEIN, A. *The World as I See It*. John Lane: London, 1935.
92. EINSTEIN, A. *Ideas and Opinions*. Crown: New York, 1954.
93. EINSTEIN, A. *The Observer*: 5 April 1954.
94. EINSTEIN, A. *Ideas and Opinions*. Dell: New York, 1973.
95. ELDER, S. T. et al. "Apparatus and procedure for training subjects to control their blood pressure", *Psychophysiology* 14: 68, 1977.
96. ELIADE, M. *Shamanism*. Pantheon: New York, 1964.
97. ELKIN, A. P. *The Australian Aborigines*. Angus & Robertson: Sydney, 1942.
98. ELKIN, A. P. *Aboriginal Men of High Degree*. St. Martins: New York, 1977.
99. ELLIOT, J. S. *Dowsing: One Man's Way*. Jersey, 1977.
100. ELLIS, K. *Number Power*. St. Martins: New York, 1978.
101. EMERSON, J. N. "Intuitive archaeology", *The Midden* 5: 3, 1973.
102. EMERSON, J. N. "Intuitive archaeology: a psychic approach", *New Horizons* 1: 14, 1974.
103. EMLEN, S. T. "The stellar-orientation system of a migratory bird", *Scientific American* 233: 102, 1975.
104. ENGEL, B. T. "Operant conditioning of cardiac function", *Psychophysiology* 9: 161, 1972.
105. EVANS-PRITCHARD, E. E. *Witchcraft, Oracles and Magic Among the Azande*. Oxford University Press: London, 1937.
106. FARADAY, M. "Experimental researches in electricity", *Philosophical Transaction of the Royal Society of London* 122: 125, 1832.
107. FARB, P. *Word Play*. Jonathan Cape: London, 1973.
108. FARMER, E. & CHAMBERS, E. "A psychological study of individual differences in accident rates", *Report of the Industrial Fatigue Research Board* 38: London, 1926.
109. FEILDING, E. *Sittings with Eusapia Palladino*. University Books: New York, 1963.
110. FERGUSON, M. *The Brain Revolution*. Davis-Poynter: London, 1974.
111. FERGUSON, S. Personal communication, 1984.
112. FIRTH, R. *Tikopia Ritual and Belief*. Beacon: Boston, 1967.
113. FODOR, N. "The poltergeist psychoanalysed", *Psychiatric Quarterly* 22: 195, 1948.
114. FODOR, N. *Between Two Worlds*, Parker: New York, 1964.

115. FORESTER, T. (ed.) *The Microelectronics Revolution*. Blackwell: Oxford, 1980.
116. FOSTER, A. "ESP tests with American Indian children", *Journal of Parapsychology* 7: 94, 1943.
117. FOSTER, D. *The Intelligent Universe*. Abelard: London, 1975.
118. FOX, M. *Understanding Your Cat*. Random House: New York, 1974.
119. FROST, W. H. "The age selection of mortality from tuberculosis . . .", *American Journal of Hygiene* 30: 31, 1939.
120. FULLER, J. G. *The Airmen Who Would Not Die*. Putnam: New York, 1979.
121. FUTUYAMA, D. J. *Evolutionary Biology*. Sinauer: Sunderland, Massachusetts, 1979.
122. GALILEO, G. "Sopra le scoperte dei dadi", *Opera* 8: 591, 1898.
123. GARDNER, M. *Science: Good, Bad and Bogus*. Oxford University Press: Oxford, 1983.
124. GAULD, A. *Mediumship and Survival*. Heinemann: London, 1982.
125. GAULD, A. & CORNELL, A. D. *Poltergeists*. Routledge and Kegan Paul: London, 1979.
126. GOETHE, J. *The Sorrows of Werther*. 1774.
127. GOLDBERG, P. *The Intuitive Edge*. Turnstone: Wellingborough, 1985.
128. GOODMAN, J. *Psychic Archaeology*. Putnam: New York, 1977.
129. GOSS, M. *The Evidence for Phantom Hitchhikers*. Aquarian: Wellingborough, 1984.
130. GOULD, L. L. "Formation Flight in the Canada Goose", Masters thesis at the University of Rhode Island: 1972.
131. GOULD, S. J. *The Panda's Thumb*. Norton: New York, 1980.
132. GRATTAN-GUINNESS, I. (ed.) *Psychical Research*. Aquarian: Wellingborough, 1982.
133. GRAVES, T. *Dowsing*. Turnstone: London, 1976.
134. GREEN, C. *Out-of-the-Body Experiences*. Hamish Hamilton: London, 1968.
135. GREEN, E. E. et al. "Feedback technique for deep relaxation", *Psychophysiology* 6: 371, 1969.
136. GREEN, E. E. et al. "Voluntary control of internal states", *Journal of Transpersonal Psychology* 2: 1, 1970.
137. GREGORY, A. In Introduction to VASILIEV (Ref. 384).
138. GRIBBIN, J. *In Search of Schrödinger's Cat*. Wildwood: London, 1984.
139. GRIGG, E. R. N. "The arcana of tuberculosis", *American Review of Tuberculosis* 78: 151, 1958.
140. GRIMBLE, A. *Migrations, Myth and Magic from the Gilbert Islands*. Routledge & Kegan Paul: London, 1972.

141. GUILLEN, M. A. *Bridges to Infinity*. Houghton Mifflin: New York, 1983.
142. GURNEY, E. et al. *Phantasms of the Living*. Trubner: London, 1886.
143. HALL, R. *Animals Are Equal*. Wildwood: London, 1980.
144. HAMILTON, W. D. "The genetical theory of social behavior", *Journal of Theoretical Biology* 7: 1, 1964.
145. HANSEN, J. "Can science allow miracles?", *New Scientist* 8 April 1973.
146. HARALDSSON, E. & OSIS, K. "The appearance and disappearance of objects . . .", *Journal of the American Society for Psychical Research* 71: 33, 1977.
147. HARDY, A. *The Living Stream*. Collins: London, 1965.
148. HARDY, A. *The Biology of God*. Jonathan Cape: London, 1975.
149. HARDY, A. *The Spiritual Nature of Man*. Oxford University Press: London, 1981.
150. HARDY, A. et al. *The Challenge of Chance*. Hutchinson: London, 1973.
151. HARNER, M. *The Way of the Shaman*. Harper & Row: New York, 1980.
152. HARNER, M. (ed.) *Hallucinogens and Shamanism*. Oxford University Press: New York, 1973.
153. HARRIS, M. *Cows, Pigs, Wars and Witches*. Random House: New York, 1974.
154. HART, H. "ESP projection", *Journal of the American Society for Psychical Research* 48: 121, 1954.
155. HARTWELL, J. et al. "A study of physiological variables associated with out-of-body experiences", in MORRIS (Ref. 251).
156. HARVIE, R. "Probability and serendipity", in HARDY (Ref. 150).
157. HASTED, J. *The Metal-Benders*. Routledge & Kegan Paul: London, 1981.
158. HAWKINS, G. S. & WHITE, J. B. *Stonehenge Decoded*. Souvenir: London, 1965.
159. HAY, D. *Exploring Inner Space*. Penguin: Harmondsworth, 1982.
160. HAYAKAWA, S. I. *Language in Thought and Action*. Allen & Unwin: London, 1952.
161. HEDIN, P. A. (ed.) *Plant Resistance to Insects*. American Chemical Society, Symposium Series 208, 1983.
162. HEGGIE, D. C. *Megalithic Science*. Thames & Hudson: London, 1981.
163. HEPPER, P. G. "Sibling recognition in the rat", *Animal Behaviour* 31: 1177, 1983.
164. HEPPNER, F. H. "Avian flight formations", *Bird-Banding* 45: 160, 1974.

165. HILGARD, E. R. *Hypnotic Susceptibility*. Harcourt Brace: New York, 1965.
166. HOFFMAN, J. G. *The Life and Death of Cells*. Hutchinson: London, 1958.
167. HOLDEN, A. & SINGER, P. *Crystals and Crystal-gazing*. Heinemann: London, 1961.
168. HONORTON, C. & BARKSDALE, W. "PK performance . . .", *Journal of the American Society for Psychical Research* 66: 208, 1972.
169. HOROWITZ, K. A. et al. "Plant primary perception . . .", *Science* 189: 478, 1975.
170. HOYLE, F. *Galaxies, Nuclei and Quasars*. Harper & Row: New York, 1964.
171. HOYLE, F. *The Intelligent Universe*. Michael Joseph: London, 1983.
172. HUBEL, D. H. "Attention units in the auditory cortex", *Science* 129: 1279, 1959.
173. HUGHES, R. "The great leap for mankind", *Sunday Telegraph*: 5 August 1984.
174. HUME, E. *Essays*. London, 1875.
175. HUMPHREY, N. "Consciousness: a just-so story", *New Scientist*: 19 August 1982.
176. HUNT, M. *The Universe Within*. Simon & Schuster: New York, 1982.
177. HUNTER, I. M. L. "An exceptional talent for calculative thinking", *British Journal of Psychology* 53: 243, 1962.
178. HUXLEY, A. *The Devils of Loudun*. Chatto & Windus: London, 1952.
179. HUXLEY, A. *The Doors of Perception*. Chatto & Windus: London, 1954.
180. HUXLEY, T. H. *Collected Essays*. London, 1881.
181. IAMBLICHUS *The Mysteries*. London, 1895.
182. IKEMA, Y. & NAKAGAWA, S. "A psychosomatic study of contagious dermatitis", *Kyushu Journal of Medical Science* 13: 335, 1962.
183. INGLIS, B. *Natural and Supernatural*. Hodder & Stoughton: London, 1977.
184. INGLIS, B. "Hysteria", *The Times*: 12 April 1982.
185. INGLIS, B. *The Paranormal*. Granada: London, 1985.
186. INGLIS, B. & WEST, R. *The Alternative Health Guide*. Michael Joseph: London, 1983.
187. ISAACS, J. D. "A mass screening technique for locating PKMB agents", *Psychoenergetics* 4: 125, 1981.
188. ISAACS, J. D. "Psychokinesis", in GRATTAN-GUINNESS (Ref. 132).

276

189. ISAACS, J. D. "Dowsing", in GRATTAN-GUINNESS (Ref. 132).
190. JAMES, W. *Varieties of Religious Experience*. New English Library: New York, 1958.
191. JANET, P. *The Mental State of Hystericals*. Putnam: New York, 1901.
192. JANTSCH, E. *The Self-Organizing Universe*. Pergamon: Oxford, 1980.
193. JEANS, J. *The Mysterious Universe*. University Press: Cambridge, 1931.
194. JENNESS, D. "The Carrier Indians of the Bulkley River", *Bulletin of the Bureau of American Ethnology* 133, 1943.
195. JOHANSON, D. C. & EDEY, M. A. *Lucy*. Granada: London, 1981.
196. JOHNSON, R. *The Imprisoned Splendour*. Harper & Row: New York, 1953.
197. JOHNSON, R. *Watchers on the Hills*. Harper & Row: New York, 1959.
198. JOHNSON, R. F. Q. & BARBER, T. X. "Hypnotic suggestions for blister formation", *American Journal of Clinical Hypnosis* 18: 172, 1976.
199. JOHNSON, R. F. Q. & BARBER, T: X. "Hypnosis, suggestion and warts", *American Journal of Clinical Hypnosis* 20: 165, 1978.
200. JUNG, C. G. *Synchronicity*. Routledge & Kegan Paul: London, 1972.
201. KALMIJN, A. J. "The electrical sense of sharks and rays", *Journal of Experimental Biology* 55: 371, 1971.
202. KALMIJN, A. J. "The electric and magnetic sense of sharks, skates and rays", *Oceanus* 20: 45, 1977.
203. KAMMERER, P. *Das Gesetz der Serie*. DVA: Stuttgart, 1919.
204. KARGER, F. & ZICHA, G. "Physical investigation of psychokinetic phenomena in Rosenheim . . .", *Proceedings of the Parapsychological Association* 5: 33, 1968.
205. KEATS, J. Letter to Benjamin Bailey: 1817.
206. KEYES, D. *The Minds of Billy Milligan*. Random House: New York, 1981.
207. KLINE, M. *Mathematics: The Loss of Certainty*. Oxford University Press: New York, 1980.
208. KLUCKHOHN, C. "Navaho witchcraft", *Papers of the Peabody Museum* 24, 1944.
209. KOESTLER, A. *The Invisible Writing*. Hutchinson: London, 1954.
210. KOESTLER, A. *The Ghost in the Machine*. Hutchinson: London, 1967.

211. KOESTLER, A. *The Case of the Midwife Toad*. Hutchinson: London, 1971.
212. KOESTLER, A. "Speculations on problems beyond our present understanding", in HARDY (Ref. 150).
213. KOESTLER, A. *Sunday Times*: 5 May 1974.
214. KOESTLER, A. *Janus*. Hutchinson: London, 1978.
215. KOHLER, W. *The Mentality of Apes*. Kegan Paul: London, 1927.
216. KRIPPNER, S. (ed.) Advances in Parapsychological Research: *Psychokinesis*. Plenum: New York, 1977.
217. LAFONTAINE, C. *Mémoires d'un Magnétiseur*. Paris, 1866.
218. LAMARCK, J. P. *Philosophique Zoologique*. Paris, 1873.
219. LANG, A. *Magic and Religion*. London, 1901.
220. LANG, P. J. et al. "Effect of feedback and instructional set . . .", *Journal of Experimental Psychology* 75: 425, 1967.
221. LAUBSCHER, B. J. F. *Sex Custom and Psychopathology*. McBride: New York, 1938.
222. LAWDEN, D. F. "Psychical research physics", In GRATTAN-GUINNESS (Ref. 132).
223. LEAHEY, T. H. & LEAHEY, G. E. *Psychology's Occult Doubles*. Nelson-Hall: Chicago, 1981.
224. LE SHAN, L. *Clairvoyant Reality*. Turnstone: London, 1974.
225. LE SHAN, L. *From Newton to ESP*. Turnstone: London, 1984.
226. LETTVIN, J. Y. et al. "What the frog's eye tells the frog's brain", *Proceedings of the Institute of Radio Engineers* 47: 140, 1959.
227. LEWIS, A. O. *Of Men and Machines*. Dutton: New York, 1963.
228. LEWIS, I. M. *Ecstatic Religion*. Penguin: Harmondsworth, 1971.
229. LISSAMAN, P. B. S. & SHOLLENBERGER, C. A. "Formation flight of birds", *Science* 168: 1003, 1970.
230. LOCKYER, J. N. *The Dawn of Astronomy*. Macmillan: London, 1894.
231. LOVELOCK, J. E. *Gaia*. Oxford University Press: London, 1979.
232. LUSCHER, M. "Air-conditioned termite nests", *Scientific American* 205: 138, 1961.
233. MABBETT, I. W. "Defining the paranormal", *Journal of Parapsychology* 46: 337, 1982.
234. MANNING, M. *In the Minds of Millions*. Allen: London, 1977.
235. MARAIS, E. *The Soul of the Ape*. Blond: London, 1969.
236. MARAIS, E. *The Soul of the White Ant*. Cape: London, 1971.
237. MARINO, A. A. & BECKER, R. O. "High voltage lines", *Environment* 20: 6, 1978.
238. MASLOW, A. *Religious Values and Peak Experiences*. State University Press: Ohio, 1962.
239. MAUSKOPF, S. H. & McVAUGH, M. R. *The Elusive Science*. Johns Hopkins University: Baltimore, 1981.

240. MAY, R. *The Courage to Create*. Norton: New York, 1975.
241. MERMET, A. *Principles and Practice of Radiesthesia*. London, 1935.
242. MICHELL, J. *A Little History of Astro-archaeology*. Thames & Hudson: London, 1977.
243. MICHELL, J. & RICKARD, R. J. M. *Phenomena*. Thames & Hudson: London, 1977.
244. MICHELL, J. & RICKARD, R. J. M. *Living Wonders*. Thames & Hudson: London, 1982.
245. MIDDLETON, J. (ed.) *Magic, Witchcraft and Curing*. University of Texas Press: Austin, 1967.
246. MILLER, N. "Learning of visceral and glandular responses", *Science* 163: 434, 1969.
247. MITCHELL, E. *Psychic Exploration*. Putnam: New York, 1974.
248. MONOD, J. *Chance and Necessity*. Collins: London, 1972.
249. MONROE, R. A. *Journeys Out of the Body*. Doubleday: New York, 1971.
250. MOORE-EDE, M. C. et al. *The Clocks That Time Us*. Harvard University Press: Cambridge, 1982.
251. MORRIS, J. D. et al. (eds.) *Research in Parapsychology 1974*. Scarecrow: New Jersey, 1975.
252. MORRIS, R. L. "An experimental approach to the survival problem", *Theta* 33: 34, 1971.
253. MORRIS, R. L. "The Use of detectors for out-of-body experiences", in ROLL (Ref. 313).
254. MORSE, D. et al. "Tuberculosis in Ancient Egypt", *American Review of Tuberculosis* 90: 524, 1964.
255. MULDOON, S. & CARRINGTON, H. *The Phenomena of Astral Projection*. Rider: London, 1951.
256. MURPHET, H. *Sai Baba: Man of Miracles*. Muller: London, 1973.
257. MURRAY, R. W. "The response of the ampullae of Lorenzini . . .", *Journal of Experimental Biology* 39: 119, 1962.
258. MYERS, F. *Human Personality*. Longmans: London, 1903.
259. McAULIFFE, K. "The mind fields", *Omni*: August 1984.
260. McCLINTOCK, M. K. "Menstrual synchrony and suppression", *Nature* 229: 244, 1971.
261. McCONNELL, R. A. "Remote night tests for PK", *Journal of the American Society for Psychical Research* 49: 99, 1955.
262. MACKAY, C. *Extraordinary Popular Delusions*. National Illustrated Library: London, 1852.
263. McKENZIE, J. H. "The haunted millgirl", *Quarterly Transactions of the British College of Psychic Science* 182, 1925.
264. MACLAY, G. R. *A Short History of the Idea that a Human Society is a Living Organism*. Unpublished MS, 1983.

265. McWHIRTER, N. D. *Guinness Book of Records*. Guinness: Enfield, 1984.

266. NASH, C. B. & NASH, C. S. "Physical and metaphysical parapsychology", *Journal of Parapsychology* 27: 283, 1963.

267. NEHER, A. "A physiological explanation of unusual behaviour in ceremonies involving drums", *Human Biology* 34: 151, 1962.

268. NEWTON, I. *Mathematical Principles of Natural History*. London, 1687.

269. NOBLE, G. K. & JAECKLE, M. E. in TAYLOR (Ref. 368).

270. NOONAN, H. M. "Science and the Psychical Research Movement". Doctoral thesis at the University of Pennsylvania, 1977.

271. ORNSTEIN, R. E. *The Psychology of Consciousness*. Freeman: San Francisco, 1972.

272. OSIS, K. "Perceptual experiments on out-of-body experience", in MORRIS (Ref. 251).

273. OWEN, I. M. & SPARROW, M. *Conjuring up Philip*. Fitzhenry: Toronto, 1976.

274. OYSTEIN, O. "Pascal and the invention of probability theory", *American Mathematics Monthly* 67: 409, 1960.

275. PAGENSTECHER, G. "Past events seership", *Proceedings of the American Society for Psychical Research* 16: 1, 1922.

276. PELTON, R. *The Devil and Karen Kingston*. Pocket Books: New York, 1977.

277. PODMORE, F. *From Mesmer to Christian Science*. University Books: New York, 1909.

278. PONIATOWSKI, S. "Parapsychological probing of prehistoric cultures", in GOODMAN (Ref. 128).

279. POPE, D. "ESP tests with primitive people", *Parapsychology Bulletin* 30: 1, 1953.

280. POPPER, K. *The Logic of Scientific Discovery*. Basic Books: New York, 1959.

281. POPPER, K. *Objective Knowledge*. Oxford University Press: London, 1972.

282. PORTER, R. H. & MOORE, J. D. "Human kin recognition by olfactory cues", *Physiology and Behavior* 27: 493, 1981.

283. POYNTON, J. C. "Results of an out-of-body survey", in POYNTON (Ref. 284).

284. POYNTON, J. C. (ed.) *Parapsychology in South Africa*. SASPR: Johannesburg, 1975.

285. PRATT, J. G. & ROLL, W. G. "The Seaford disturbances", *Journal of Parapsychology* 22: 79, 1958.

286. PREISER, F. E. *Environmental Design Research*. Hutchinson & Ross: Stroudsburg, 1973.

287. PRIBRAM, K. H. "The neurophysiology of remembering", *Scientific American* 228: 73, 1969.

288. PRICE, H. H. "Some philosophical questions about telepathy . . .", *Philosophy* 15: 363, 1940.

289. PRINCE, W. F. *The Enchanted Boundary*. Little, Brown: Boston, 1930.

290. PRZIBRAM, H. "Paul Kammerer als Biologe", *Monistische Monatshefte* 401, 1926.

291. PULOS, L. "Mesmerism revisited", *American Journal of Clinical Hypnosis* 22: 206, 1980.

292. PUTNAM, F. "Traces of Eve's Faces", *Psychology Today*: October 1982.

293. RADIN, P. *Primitive Man as Philosopher*. Dover: New York, 1957.

294. RANDALL, J. L. *Parapsychology and the Nature of Life*. Souvenir: London, 1975.

295. RANDALL, J. L. *Psychokinesis*. Souvenir: London, 1982.

296. RAO, K. N. *Textbook of Tuberculosis*. Vikas: New Delhi, 1981.

297. REBER, A. In *Psychology Today*: August, 1984.

298. REES, M. J. et al. *Black Holes, Gravitational Waves and Cosmology*. Gordon & Breach: New York, 1974.

299. RHINE, J. B. *Extra-sensory Perception*. SPR: Boston, 1934.

300. RHINE, J. B. "Location of hidden objects by a man-dog team", *Journal of Parapsychology* 35: 18, 1971.

301. RHINE, J. B. & FEATHER, S. R. "The study of cases of psi-trailing in animals", *Journal of Parapsychology* 26: 1, 1962.

302. RHINE, L. E. *Mind Over Matter*. Macmillan: New York, 1970.

303. RHOADES, D. F. "Responses of alder and willow to attack by tent caterpillars and webworms", in HEDIN (Ref. 161).

304. RICHARDS, S. *Luck, Chance and Coincidence*. Aquarian: Wellingborough, 1985.

305. ROBINS, D. "The Dragon Project and the talking stones", *New Scientist*: 21 October 1982.

306. ROGO, D. S. "Psi and shamanism", *Parapsychology Review* 14: 5, 1983.

307. ROGO, D. S. "Searching for psi in primitive cultures", *Parapsychology Review* 15: 1, 1984.

308. ROLL, W. G. "Token object matching tests", *Journal of the American Society for Psychical Research* 60: 363, 1966.

309. ROLL, W. G. *The Poltergeist*. Scarecrow: New Jersey, 1976.

310. ROLL, W. G. "ESP and memory", in WHITE (Ref. 409).

311. ROLL, W. G. & PRATT, J. G. "The Miami disturbances", *Journal of the American Society for Psychical Research* 65: 409, 1971.

312. ROLL, W. G. et al. "Radial and tangential forces in the Miami Poltergeist", *Journal of the American Society for Psychical Research* 67: 267, 1973.

313. ROLL, W. G. et al. (eds.) *Researches in Parapsychology 1973*. Scarecrow: New Jersey, 1974.

314. ROSE, J. (ed.) *Proceedings of the International Congress of Cybernetics*. Gordon & Breech: New York, 1970.

315. ROSE, L. & ROSE, R. "Psi experiments with Australian Aborigines", *Journal of Parapsychology* 15: 122, 1951.

316. ROSE, R. "A second report on psi experiments with Australian Aborigines", *Journal of Parapsychology* 19: 92, 1955.

317. ROSE, R. *Living Magic*. Rand McNally: New York, 1956.

318. ROSE, S. *The Conscious Brain*. Penguin: Harmondsworth, 1976.

319. ROSENTHAL, G. A. et al. *Herbivores*. Academic Press: New York, 1979.

320. ROSS, W. "Archaeological map dowsing", Paper presented to the Canadian Archaeological Association Annual Meeting at Thunder Bay: March 1975.

321. RUSSELL, E. W. *Report on Radionics*. Spearman: London, 1973.

322. RUSSELL, P. *The Awakening Earth*. Routledge & Kegan Paul: London, 1982.

323. SAMUELS, M. & SAMUELS, N. *Seeing with the Mind's Eye*. Random House: New York, 1976.

324. SARGENT, C. "Exploring psi in the Ganzfeld", *Parapsychological Monographs* 17, 1980.

325. SAWYER, W. W. *Prelude to Mathematics*. Penguin: Harmondsworth, 1955.

326. SCHLITZ, M. & GRUBER, E. "Transcontinental remote viewing", *Journal of Parapsychology* 44: 305, 1981.

327. SCHMIDT, H. "A PK test with electronic equipment", *Journal of Parapsychology* 34: 175, 1970.

328. SCHMIDT, H. "The comparison of PK action on two different random number generators", *Journal of Parapsychology* 38: 47, 1974.

329. SCHMIDT, H. "Pk effect on pre-recorded targets", *Journal of the American Society for Psychical Research* 70: 267, 1976.

330. SCHNEIRLA, T. C. "Army ants", in TOPOFF (Ref. 378).

331. SCHOPENHAUER, A. *Short Philosophical Essays*. Oxford University Press: London, 1974.

332. SCHRÖDINGER, E. Inaugural address delivered at the University of Zurich in December 1922.

333. SCHRÖDINGER, E. *What Is Life?* Cambridge University Press: New York, 1967.

334. SELOUS, E. *Thought Transference (or What?) in Birds*. Constable: London, 1931.

335. SHALLIS, M. *The Silicon Idol*. Oxford University Press: Oxford, 1984.

336. SHAPIN, S. & COLY, L. (eds.) *Psi and States of Awareness.* Parapsychology Foundation: New York, 1978.
337. SHAPIRO, D. et al. "Differentiation of heart rate by operant conditioning", *Psychosomatic Medicine* 32: 417, 1970.
338. SHATTUCK, E. H. *An Experiment in Mindfulness.* Dutton: New York, 1958.
339. SHEEHAN, P. W. & PERRY, C. W. *Methodologies of Hypnosis.* Lawrence Erlbaum: New Jersey, 1976.
340. SHEILS, D. "A cross-cultural survey of beliefs in out-of-the-body experiences", *Journal of the Society for Psychical Research* 49: 697, 1978.
341. SHELDRAKE, R. *A New Science of Life.* Blond & Briggs: London, 1981.
342. SHEPHER, J. "Mate selection amongst second generation kibbutz adolescents . . .", *Archives of Sexual Behaviour* 1: 293, 1971.
343. SHERRINGTON, C. *Man on His Nature.* Cambridge University Press: Cambridge, 1951.
344. SHIRLEY, R. *The Mystery of the Human Double.* University Books: New York, 1965.
345. SILANDER, J. A. et al. *Oecologia* 58: 415, 1983.
346. SIZEMORE, C. & PITILLO, E. *I'm Eve.* Doubleday: New York, 1977.
347. SMITH, A. *The Body.* Viking: London, 1985.
348. SOCIETY FOR PSYCHICAL RESEARCH *Census of Hallucinations.* SPR: London, 1894.
349. STANFORD, R. G. "Conceptual frameworks of contemporary psi research", in WOLMAN (Ref. 419).
350. STANFORD, R. G. et al. "Psychokinesis as a psi-mediated instrumental response", *Journal of the American Society for Psychical Research* 69: 127, 1975.
351. STAPLEDON, O. *Starmaker.* Penguin: Harmondsworth, 1972.
352. STEPTOE, A. et al. "The learned control of differential temperature in the human earlobes", *Biological Psychology* 1: 237, 1974.
353. STEVENSON, I. & PASRICHA, S. "A case of secondary personality with xenoglossy", *American Journal of Psychiatry* 136: 1591, 1979.
354. STICKROD, G. et al. "In utero taste-odour aversion conditioning in the rat", *Physiological Behavior* 28: 5, 1982.
355. STROMEYER, C. F. & PSOTKA, J. "The detailed texture of eidetic images", *Nature* 225: 346, 1970.
356. STUKELEY, W. *Stonehenge.* London, 1740.
357. SZENT-GYÖRGYI, A. *Bioenergetics.* Morrow: New York, 1957.
358. TALBOT, M. *Mysticism and the New Physics.* Routledge & Kegan Paul: London, 1981.

359. TANSLEY, D. V. *Omens of Awareness*. Spearman: London, 1977.

360. TANSLEY, D. V. *Radionics: Science or Magic?* Daniel: Saffron Walden, 1982.

361. TARG, R. et al. *Research in Parapsychology 1979*. Scarecrow: New Jersey, 1980.

362. TARG, R. & HARARY, K. *The Mind Race*. Villard: New York, 1984.

363. TARG, R. & PUTHOFF, H. "Information transmission under conditions of sensory shielding", *Nature* 251: 602, 1974.

364. TARG, R. & PUTHOFF, H. *Mind Reach*. Delacorte: New York, 1977.

365. TART, C. T. *Psi*. Dutton: New York, 1977.

366. TART, C. T. "The controversy about psi", *Journal of Parapsychology* 46: 313, 1982.

367. TART, C. T. et al. *Mind at Large*. Praeger: New York, 1979.

368. TAYLOR, G. R. *The Great Evolution Mystery*. Secker & Warburg: London, 1983.

369. TAYLOR, J. G. *Superminds*. Macmillan: London, 1973.

370. TAYLOR, J. G. *Science and the Supernatural*. Temple Smith: London, 1980.

371. TAYLOR, J. G. & BALANOVSKI, E. "Can electromagnetism explain ESP?", *Nature* 275: 64, 1978.

372. THIGPEN, C. H. & CLECKLEY, H. M. *The Three Faces of Eve*. Secker & Warburg: London, 1957.

373. THOM, A. *Megalithic Sites in Britain*. Clarendon Press: Oxford, 1967.

374. THOMAS, S. "Have you ever heard a stone talking?", *The Guardian*: 25 June 1983.

375. THOMPSON, D. W. *On Growth and Form*. University Press: Oxford, 1917.

376. TINTEROW, M. M. *Foundations of Hypnosis*. Thomas: Springfield, Illinois, 1970.

377. TIZANE, E. *Sur la Piste de l'Homme Inconnu*. Amiot-Dumont: Paris, 1951.

378. TOPOFF, H. R. (ed.) *Army Ants*. Freeman: San Francisco, 1971.

379. TRAVIS, W. Personal communications, 1984.

380. ULLMAN, M. et al. *Dream Telepathy*. Turnstone: London, 1973.

381. VAN DE CASTLE, R. L. "Psi abilities in primitive groups", *Proceedings of the Parapsychological Association* 7: 97, 1970.

382. VAN DE CASTLE, R. L. "Anthropology and psychic research", in MITCHELL (Ref. 247).

383. VAN HOVEN, W. "Tree's secret warning system against browsers", *Custos* 13: 11, 1984.

284

384. VASILIEV, L. L. *Experiments in Distance Influence*. Wildwood: London, 1976.

385. VON FOERSTER, H. "On constructing a reality", in PREISER (Ref. 286).

386. VON URBAN, R. *Beyond Human Knowledge*. Pageant Press: New York, 1958.

387. WALD, G. "Life and light", *Scientific American* 201: 92, 1959.

388. WALDMAN, B. "Tadpoles have a familiar smell", *New Scientist*: 30 September 1982.

389. WALKER, E. H. "Consciousness and quantum theory", in MITCHELL (Ref. 247).

390. WALKER, M. M. et al. "A candidate magnetic sense organ in the yellowfin tuna", *Science* 224: 751, 1984.

391. WALLACE, R. K. et al. "A wakeful hypometabolic state", *American Journal of Physiology* 221: 795, 1971.

392. WALLACE, R. K. & BENSON, H. "The physiology of meditation", *Scientific American* 226: 89, 1971.

393. WALLAS, G. *The Art of Thought*. Harcourt Brace: New York, 1929.

394. WALLIS, R. (ed.) "On the margins of science", *Social Review Monograph* 27: University of Keele, 1979.

395. WATKINS, A. *The Old Straight Track*. Methuen: London, 1925.

396. WATSON, L. *Supernature*. Hodder & Stoughton: London, 1973.

397. WATSON, L. *The Romeo Error*. Hodder & Stoughton: London, 1974.

398. WATSON, L. *Lifetide*. Hodder & Stoughton: London, 1979.

399. WATSON, L. *Earthworks*. Hodder & Stoughton: London, 1986.

400. WEAVER, W. *Lady Luck: The Theory of Probability*. Dover: New York, 1963.

401. WEIL, A. *The Natural Mind*. Houghton Mifflin: New York, 1972.

402. WEINBERG, S. *Gravitation and Cosmology*. Wiley: New York, 1972.

403. WEIZENBAUM, J. "Where are we going?", in FORESTER (Ref. 115).

404. WEVER, R. A. *The Circadian System of Man*. Springer: New York, 1979.

405. WHEELER, W. M. *Ants*. Columbia University Press: New York, 1910.

406. WHEELER, W. M. "The ant colony as an organism", *Journal of Morphology* 22: 307, 1911.

407. WHITE, K. D. "Salivation: the significance of imagery in its voluntary control", *Psychophysiology* 15: 196, 1978.

408. WHITE, R. A. "The influence of experimentor motivation", in WOLMAN (Ref. 419).

285

409. WHITE, R. A. (ed.) *Surveys in Parapsychology*. Scarecrow: New Jersey, 1976.
410. WHITEHEAD, A. N. *Nature and Life*. Cambridge University Press: Cambridge, 1934.
411. WILBER, K. (ed.) *Quantum Questions*. Shambhala: Boulder, 1984.
412. WILBUR, C. *Sibyl*. Doubleday: New York, 1975.
413. WILLIAMS, J. E. "Stimulation of breast growth by hypnosis", *Journal of Sex Research* 10: 316, 1974.
414. WILLIAMSON, S. et al. (eds.) *Biomagnetism*. Plenum: New York, 1983.
415. WILLMER, E. N. *Cytology and Evolution*. Academic Press: London, 1970.
416. WILLSON, M. F. & BURLEY, N. *Mate Choice in Plants*. Princeton University Press: Guildford, 1985.
417. WILSON, E. O. *Sociobiology*. Harvard University Press: Massachusetts, 1975.
418. WILSON, E. O. *On Human Nature*. Harvard University Press: Massachusetts, 1978.
419. WOLMAN, B. B. (ed.) *Handbook of Parapsychology*. Van Nostrand: New York, 1977.
420. WOOD, J. E. *Sun, Moon and Standing Stones*. Oxford University Press: London, 1978.
421. WUILLEMIN, D. & RICHARDSON, B. "On the failure to recognize the back of one's own hand", *Perception* 11: 53, 1982.
422. YATES, F. A. *The Art of Memory*. Routledge & Kegan Paul: London, 1966.
423. YOUMANS, G. P. *Tuberculosis*. Saunders: Philadelphia, 1979.
424. YOUNG, J. Z. *A Model of the Brain*. Clarendon Press: Oxford, 1964.

INDEX

Main references are in **bold**. A page number preceded by 'q' (e.g. q95) denotes a reference to, or a quotation from, an author named in the Bibliography (which is itself alphabetically arranged) but not named with the reference on the page.

INDEX

Leviathan (Hobbes), 122
Levinson, Horace, 24–5
Lewis, A. O., q208
Lewis, I. M., q220
ley lines, 259–60
life
  as 'director' of Earth, 178,
    261–2
  elements of, 32
  'elusiveness', 91, 92
  as order, 39, 45
  reductionist theories of, **11**
  'rule-bending', 261
life-fields, 93
light, 36–7
lightning, 99–100
Lissaman, P. B. S. *and*
  Shollenberger, q35
Living Magic (Rose), 224
lizards, 94
Lockyer, Sir Joseph Norman,
  258
Lorenzini, ampullae of, 97,
  98
Lovelock, James E., 178
LSD (lysergic acid
  diethylamide), 109
luck/ill-luck, 13, **16**
Luscher, M., q123
Lydia (*in* Asia Minor), 13

Mabbett, I. W., q232
McAuliffe, K., q101, q102
McClintock, Martha K., 119
McConnell, R. A., 200
machines, **207–10**
  computers, 210–11
McKenzie, J. H., q208
Maclay, George R., 121,
  q132
McMullen, George, 246–7
McWhirter, Norris D., q90
'magic', 220–26
  'Earth magic', 254–60
magnetism/electro-
  magnetism, 29–30, **61–2**
  'animal magnetism', 92
  biomagnetism, **97–103**
  geomagnetic variation,
    61–2
  healing, 93–6
  radiation, 36, 85
  standing stones, 257
  telecommunications, 262
magnetite, 98
Mahadevan, Rajan, 90
Maimonides dream
  laboratory, 142
Manning, Matthew, 160
Marais, Eugène, q124,
  125–6, 128, 161–2
Margenau, Henry, 232

Marino, A. A. *and* Becker,
  q103
Maslow, Abraham Harold,
  112
mass hysteria, 133–6
materialisation, 183–5
mathematics, *see* number
Matthew, Patrick, 68–9
Mauskopf, S. H., *and*
  McVaugh, q229
May, Rollo, 74
Mayans, 13
meditation, 78, 107–8
megaliths, *see* stones,
  standing
meme, 65
memory, **240–2**
  abnormal, 89–90
  eidetic, 90
  latent, 157
  psychometry, 242–7
  racial, 139
Mendel, Gregor Johann, 10,
  11
Mendeleyev, Dmitri
  Ivanovich, 27–8
menstruation, 119
meristem, 53
Mermet, Abbé A., **251**, 253
Mesmer, Franz (*or* Friedrich
  Anton)/mesmerism,
  91–2, 146–7, 148, 160
metal-bending, **203–5**
  machinery, 207–11
metazoan
  defined, 44
Miami poltergeist, 196–7
Michell, J., q258
  and Rickard, q79, q135,
    q185
microwaves, 100, 101–2
Middleton, J., q221
migration
  ants, 127–8
  birds, 99, 113
Miller, Neal, 105
Milligan, Billy, 150–1, 159
Milton, John, 73–4, 208
mind, **57**, **240–1**
  as cause, 232
  of Earth, 260–3
  independent of body,
    145–6
  independent of brain, 137
  mental models, 187–8
  of universe, 263
  see also consciousness *etc*
mind fields, 97–103
minilab, 206–7
miracles, 182–9
Mitchell, Edgar, 203
Monod, Jacques, 21
Monroe, Robert A., 169

Moon
  geomagnetism, 61
  stone circles, 259
Moore-Ede, M. C. *et al*, q102
Mormons (Church of Jesus
  Christ of Latter-Day
  Saints), 226
morphogenetic fields, **62–4**,
  96
Morris, Robert L., 168,
  210–11
Morse, D. *et al*, q58
Mozart, Wolfgang Amadeus,
  29, **73**
Muldoon, Sylvan, 170
  and Carrington, q170
multiple personality, **146–52**
Murphet, Howard, 183
Murray, R. W., q97
mutation, genetic, 10–12
  'edited', 140
  random/guided, 67–71
  trees, 53–6
Myers, Frederic, 162–3
mysticism, *see*
  transcendence; religion

Nash, C. B. *and* Nash, q242
natural selection, **9**, 11
  'evolutionary excess',
    89–90
  ideas, 65
  limitations on, 39
  luck, 21
  random/predictive, **67–71**
  see also evolution
Nature (journal), 62, 164, 204
nautilus, 27
Navajo Indians, 221
Neher, A., q218
nervous systems
  autonomic/central, 104–9,
    112
  disinhibition, 109
  evolution, 103–4
  neurotransmitters, 92
  see also brain
New Guinea, 224
Newlands, John Alexander
  Reina, 27
New Science of Life, A
  (Sheldrake), 62
Newton, Sir Isaac, 23, 71–2
nitrogen, 32
nitrous oxide (laughing gas),
  108–9
Noble, G. K. *and* Jaeckle,
  q70
Noble, William, 250
Noonan, H. Molly, 228
Norman, Donald, 75
nuclear forces (weak/strong),
  29–30

292

INDEX